Twenty Mock Mathcounts Sprint Round Tests

Sam Chen and Guiling P. Chen

http://www.mymathcounts.com/index.php

Printed in the United States of America

ISBN - 13: 978-1456589141

ISBN-10: 1456589148

This book is dedicated to Mrs. Tonya McLawhorn, my Mathcounts coach at Arendell Parrott Academy.

--- SC

ACKNOWLEDGEMENTS

We like to credit and thank the following math competitions for their great math problems:

China Hope Cup Math Competition, China Hua Luo Geng Cup Math competition, China Elementary School Math Competition, China Middle School Math Competition, China Yingchun Cup Math competition, and China Zou Mei Math Competition.

We like to thank the following Mathcounts coaches who kindly reviewed the manuscripts and made corrections and comments:

Tonya McLawhorn

Arendell Parrott Academy

Mike Whitley

Broad Creek Middle School

Wendy Hopkins

The Epiprany School

Melanie Simmons

North East Elementary School

Sandra W. Lamm

Greenfield School

Sally Davis

Smyrna Elementary School

TABLE OF CONTENTS

This page is intentionally left blank.

MATHCOUNTS

■ Sprint Round Competition ■
Practice Test 1
Problems 1-30

Name

Date

DO NOT BEGIN UNTIL YOU ARE INSTRUCTED TO DO SO.

This round of the competition consists of 30 problems. You will have 40 minutes to complete the problems. You are not allowed to use calculators, books, or any other aids during this round. If you are wearing a calculator wrist watch, please give it to your proctor now. Calculations may be done on scratch paper. All answers must be complete, legible, and simplified to lowest terms. Record only final answers in the blanks in the right-hand column of the competition booklet. If you complete the problems before time is called, use the remaining time to check your answers.

Total Correct	Scorer's Initials

1. Which number is a prime number? 2001, 2004, 2005, 2007, 2011, and 2013.

2. How many obtuse angles are there among the four angles of 89°, 126°, 180°, and 216°?

3. How many of the numbers from 1 to 2011 are simultaneously divisible by 2, 3, and 5?

4. Alex and Bob are standing in front of a big mirror. Alex looks at the mirror and reads the number on Bob's sportswear as shown in the figure. What is the number of Bob's sportswear?

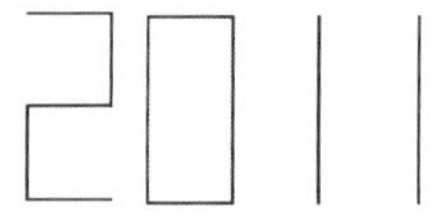

5. How many three-digit positive integers are there such that the three digits are in increasing order (like145) or decreasing order (like 321)?

6. In the sequence $7^1,\ 7^2,\ 7^3,\ \ldots,\ 7^{2011}$, how many terms have 3 as the units digit?

7. How many integer values of x if the expression $\dfrac{6x+3}{2x-1}$ represents integers?

8. Ten congruent cubes are put together as shown in the figure. If the cube A is removed, what is the difference of the surface areas before removing the cube A and after removing the cube A?

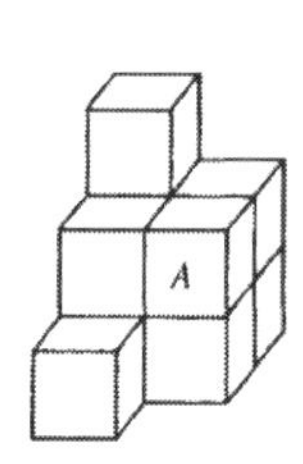

9. According to the pattern below, the positive integer 2011 is in the nth row and mth column. What is the value of $n + m$?

1	3	5	7
15	13	11	9
17	19	21	23
31	29	27	25
⋮	⋮	⋮	⋮

10. a_1, a_2, a_3, and a_4 are the number of triangles in Figure (1), (2), (3) and (4), respectively. $a_1=3$. $a_2=8$, $a_3=15$. $a_3=24$. What is a_9 ?

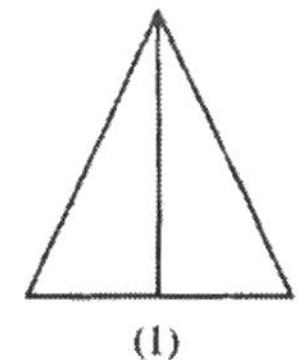
(1)

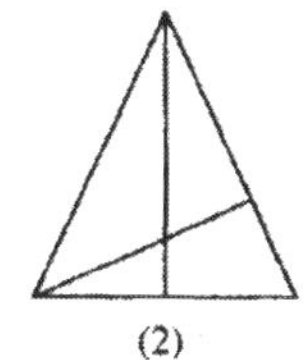
(2)

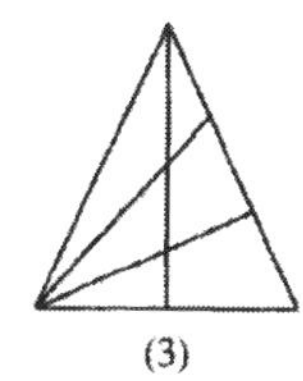
(3)

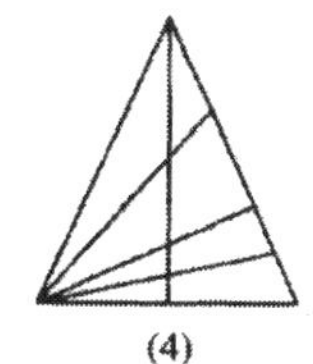
(4)

11. 2011 students are standing in a row. They are called in the following pattern: 1, 2, 3, 4, 3, 2, 1, 2, 3, 4, 3, 2, ….. What number is the 2011th student called?

12. At 3:30 P.M, the acute angle formed by the hour and the minute hands on a clock is $x°$. Find the value of x.

13. a and b are prime numbers. $3a + 2b$ is a prime number less than 20. How many pairs of (a, b) are there?

14. A number machine is shown as follows. When input $x = 3$ and $y = -2$, what is the result?

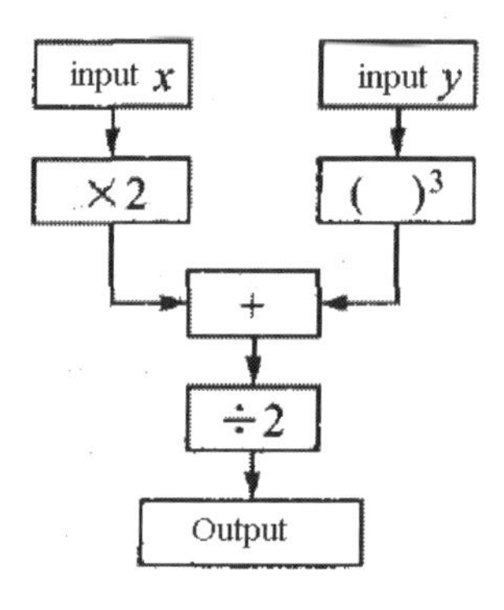

15. Four distinct balls are randomly put into three distinguishable boxes. Find the probability that no box is empty. Express your answer as a common fraction.

16. Randomly pick three digits from 1, 2, 3, 4 and 5 to form a three-digit number. Find the probability that the sum of the digits of the three-digit number is nine. Repetition of digits is allowed. Express your answer as a common fraction.

17. As shown in the figure, one toy face is obtained by sliding the other. The coordinates of the eyes of the left toy face are (-4, 2) and (-2, 2), respectively. One of the eyes of the right toy face is (3, 4). Find the sum of the coordinates of the other eye of the toy face on the right.

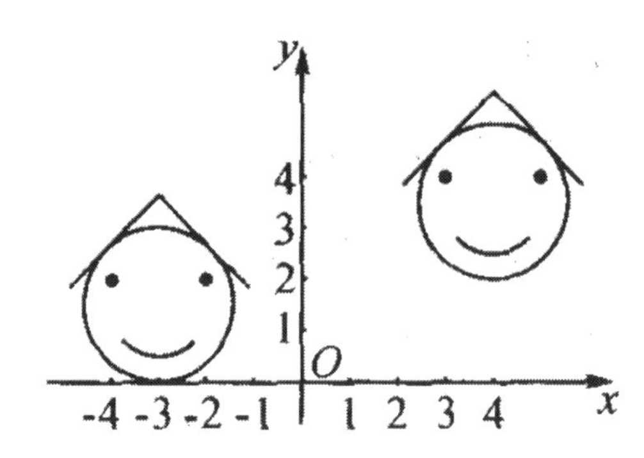

18. Less than forty identical small balls can be arranged to form a square. These balls can also be arranged to form an equilateral triangle. Find the number of small balls.

19. Add a single term to a polynomial $4x^2+1$ to result in a square number. How many different such single terms are there?

20. Alex wants to make a cylinder shaped container by using a piece of rectangle shaped metal sheet as shown in the figure. What is the volume of this container in terms of π?

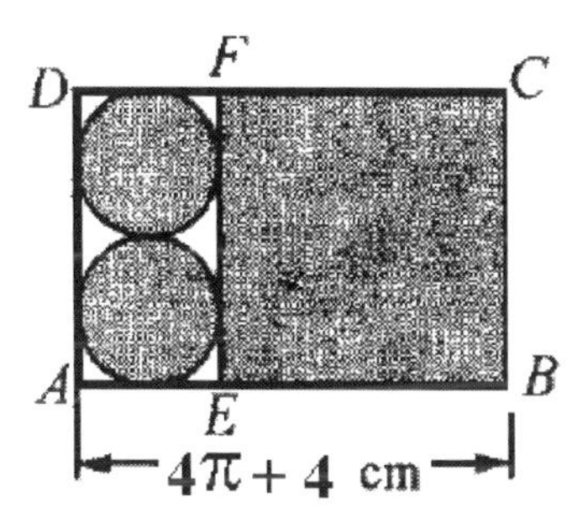

21. As shown in the figure, Both ABCD and CEFG are squares. Connect AG and AG meets CE at H. Connect HF. Find the shaded area.

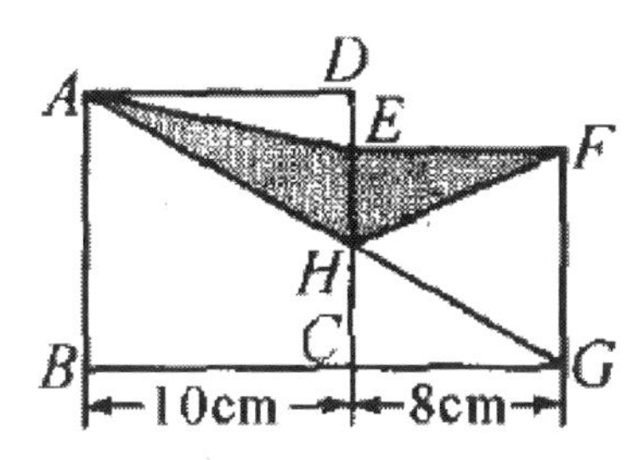

22. As shown in the figure, EABCD is a road along the lake. When Alex starts to walk along the road, he goes from west to east (from E to A). He first turn 100° at A, that is $\angle A = 100°$. Second time he turns 150° at B, that is $\angle B = 150°$. Third time he turns, at C, he walks from west toward east again. Find degrees he turns at C, that is find $\angle C$.

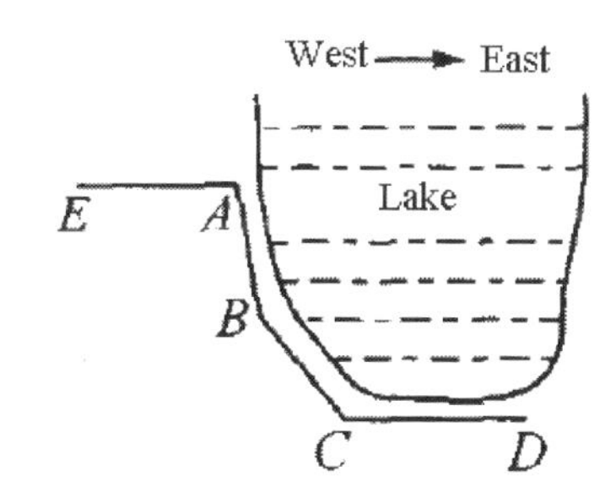

23. Draw a line L passing through point (1, 3). Line L intersects the x-axis at a and the y-axis at b. How many such lines can be drawn? Both a and b are positive integers.

24. After a math test, Ms. Leighton calculated the average score of 20 students who took the test and she got 85 points. Alex was late for the test and he made up the test later and he got exactly 85 points on the test. When Ms. Leighton re-calculated the class average, what did she get?

25. A six –digit number $\overline{a2011b}$ is divisible by 12. How many such six-digit numbers are there?

26. The sum of n interior angles and one exterior angle of a convex n-gon is 1350°. Find n.

27. There are four switches on an instrument board. If any two neighboring switches cannot be turned off at the same time, how many different combinations are there?

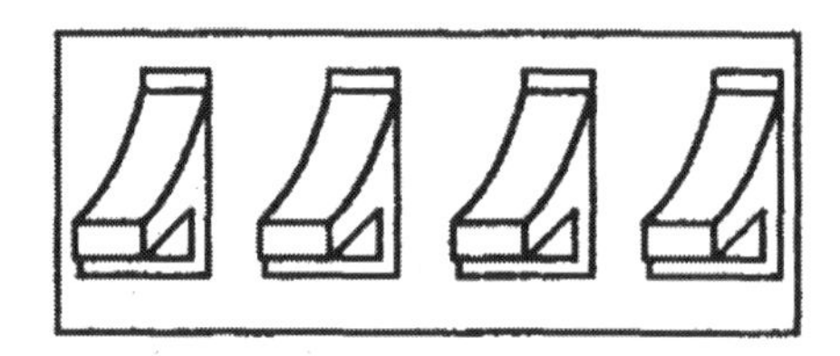

28. There are 100 students in 8th grade at Parrott Academy. 75 of them got perfect scores on Math and 80 of them got perfect scores on English. At least x students got perfect scores on both subjects and at most y students got perfect scores on both subjects. Find the value $x+y$.

29. Find a point P that is on the same plane as the equilateral triangle ABC such that ΔPAB, ΔPAC, and ΔPBC are all isosceles triangles. How many possible such points of P are there?

30. A positive number has a decimal part and an integer part. The decimal part, the integer part, and the positive number itself form a geometric sequence. Find this positive number.

Answer keys to practice test 1

1. 2011
2. 1
3. 67
4. 1105
5. 204
6. 503 (Pattern: 7, 9, 3, 1. Every 4 number has one. 2011= 502 ×4 +3)
7. 4
8. 0
9. 255. (Pattern: every 8 numbers. 2011 = 251 ×8 +3)
10. 99
11. 1 (Pattern: every 6 numbers. 2011 = 335 ×6 +1)
12. 75°
13. 3.
14. -1
15. 4/9
16. 19/125 $(P = \frac{6+3+3+6+1}{125} = \frac{19}{125})$
17. 9
18. 36
19. 5 $(-4x^2,\ -1,\ 4x,\ -4x,\ 4x^4)$.
20. 32π
21. 32
22. 130
23. 2
24. 85
25. 6 (a = 3, 6, 9 and b = 2; a= 2, 5, and 8 and b = 6)
26. 9
27. 8
28. 130 (75 + 55)
29. 10

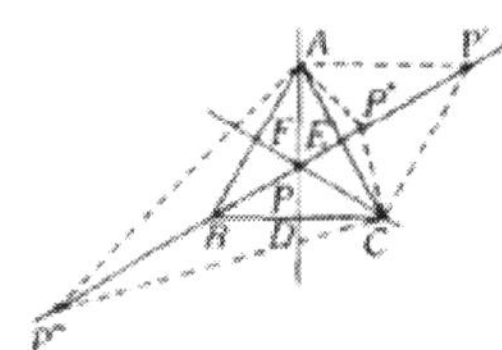

30. $\frac{\sqrt{5}+1}{2}$ $(a^2 = b(b+a) \Rightarrow (\frac{b}{a})^2 - \frac{b}{a} - 1 = 0 \Rightarrow \frac{a}{b} = \frac{\sqrt{5}+1}{2} \Rightarrow a = 1, b = \frac{\sqrt{5}-1}{2}$

MATHCOUNTS

■ Sprint Round Competition ■
Practice Test 2
Problems 1-30

Name

Date

DO NOT BEGIN UNTIL YOU ARE INSTRUCTED TO DO SO.

This round of the competition consists of 30 problems. You will have 40 minutes to complete the problems. You are not allowed to use calculators, books, or any other aids during this round. If you are wearing a calculator wrist watch, please give it to your proctor now. Calculations may be done on scratch paper. All answers must be complete, legible, and simplified to lowest terms. Record only final answers in the blanks in the right-hand column of the competition booklet. If you complete the problems before time is called, use the remaining time to check your answers.

Total Correct	Scorer's Initials

1. Calculate: $\left(5\frac{2}{7}+4\frac{1}{9}\right)\div\left(1\frac{1}{9}+1\frac{3}{7}\right)$.

2. A letter band folded once is shown in the figure, what is the letter? (It is not L).

3. $19+28+37+46+55+64+73+82+91+x=550$. Find the value of x.

4. Fill each square with a digit from the eight digits 0, 1, 2, 3, 4, 6, 7, 8 to make the subtraction true. What is the number subtracted?

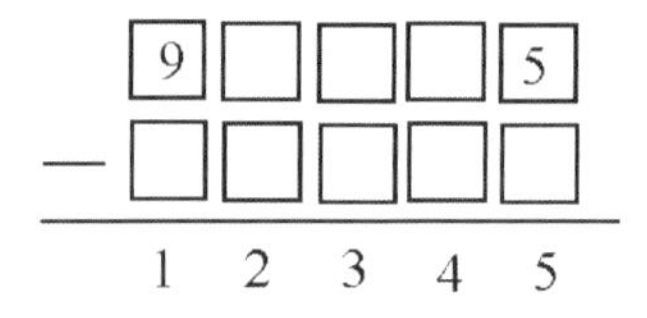

5. There are some marbles in a bag. Alex takes out half of them each time and then puts one marble back. After seven times, there are 3 marbles left in the bag. How many marbles were there in the bag originally?

6. A fifteen-story building has an elevator from the 4th floor up (including the 4th floor) and does not have an elevator from the first floor to third floor. Catherine wants to get to the top floor. It takes her 30 seconds to get to the second floor from the first floor. The elevator's speed is 10 times Catherine's speed. How many seconds does it take for her to get to the top floor if her walking speed is constant?

7. Fifteen unit cubes on the table are painted red on all visible faces. When these cubes are dissembled, what is the sum of the faces that are painted red?

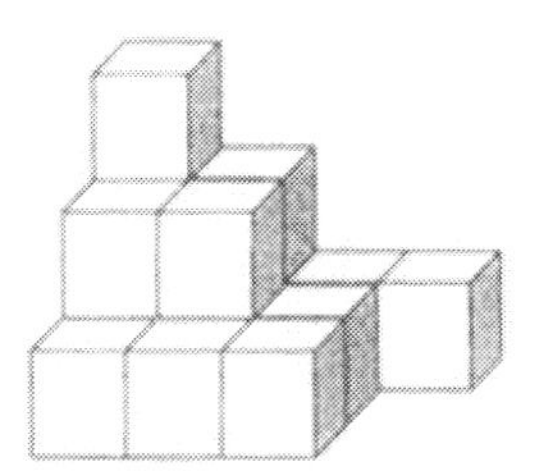

8. A rectangular paper of 48 cm long and 2 cm wide is folded in half lengthwise. The resulting figure is folded again the same way and then it is cut along the length every 2 cm. How many squares can be obtained?

9. Alex's great-grandfather Bob is over 100 years old. Bob was born in the year that is a 4-digit number with two neighboring digits identical. The sum of the 4 digits is 24. What year was Bob born?

10. Many triangles can be formed by connecting three vertices of a regular hexagon. How many of these triangles have at least one side that is a side of the hexagon?

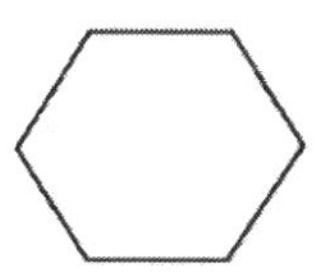

11. A small frog is jumping up a set of stairs. Each time, the frog jumps 4 cm high starting from height of 10 cm above the ground. When the frog stops after jumping 99 times, how high is the frog from the ground?

12. Karen writes down her mother's age then write her own age after her mother's age to form a 4-digit number. If the positive difference of their ages is subtracted from this 4-digit number, another 4-digit number 4289 is obtained. How old is Karen?

13. When 1 is added to a counting number, the result is a multiple of 2. When 1 is added to the double of the number, the result is a multiple of 3. When 1 is added to the triple of the number, the result is a multiple of 5. What is the smallest such number?

14. In a classroom, the number of boys is two times of the number of girls. If 12 boys walk away, then the number of girls is two times of the number of boys left. How many total students are there in this classroom?

15. What is the least number of squares that you should shade such that there is exactly one square shaded in each row, each column, and each diagonal? Two of the squares are shaded for you already.

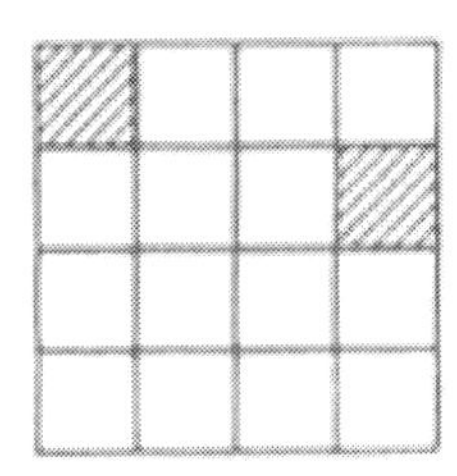

16. Only one student can drive among Alex, Bob, Catherine, and Donna. Alex says: "I can drive". Bob says: "I can't drive". Catherine says: "Alex can't drive". Donna says nothing. It is known that only one person is correct. Who can drive?

17 Define $x \otimes y = 3x + 7y$, find $(1 \otimes 1) + (2 \otimes 2) + (3 \otimes 3) + ... + (10 \otimes 10)$

18. Alex has five pieces of square-shaped chocolate with the side lengths 10 cm, 11 cm, 12 cm, 13 cm, and 14 cm, respectively. Alex will eat 2 cm^2 of them each day. How many days can these five pieces of chocolate last?

19. The average value of some distinct positive integers is 100. One of these numbers is 108. If the number 108 is taken away, the average value becomes 99. What is the greatest value of the largest number in them?

20. As shown in the figure, ABCD is a trapezoid. The areas of ΔABE and ΔADE are 2 cm^2 and 3 cm^2, respectively. Find the area of ΔCDE. Express your answer as a common fraction.

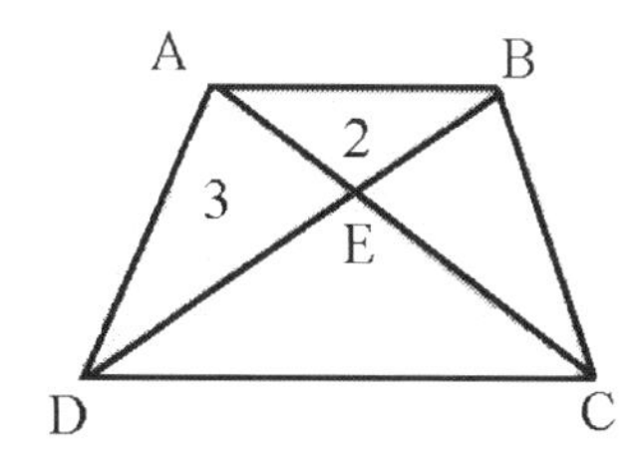

21. Take any 10 numbers from 1 to 20 and add them together to get the sum m. Add the other 10 numbers to get the sum n. How many different products of m and n are there?

22. Train A is 120 meters long with a speed of 80 km/ hour and is driving east. Train B is 280 meters long and is driving west and meets train A at the west end of a 130 meter long bridge. The two trains are separated at the east end of the bridge. What is the train B's speed?

23. Sam starts to drive from T and passes A, B, C, D, and E each once and back to T. No road can be traced twice. The numbers in the figure are the number of hours needed to cover the distance of the section. At least how many hours are needed?

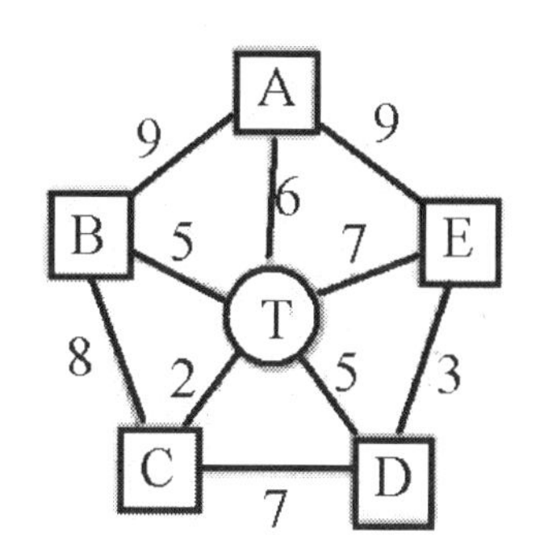

24. A TV store makes a profit of $60 per TV sold at the original price. This week the price is marked down. The number of TVs sold is doubled and the total profit increased by 0.5 times. How many dollars are each TV's price reduced?

25. List the numbers from 2010 to 1020 from the largest to the smallest in a row to form a positive integer: 2010200920082007.....10211020. What is the 999th digit counted from the left end?

26. The sum of 50 distinct positive integers is 2010. At most how many even positive integers are there?

27. As shown in the figure, ABC is a triangle with the area 1. Extend BA to D such that DA = AB. Extend CA to E such that EA = 2AC. Extend CB to F such that FB = 3BC. Find the area of triangle DEF.

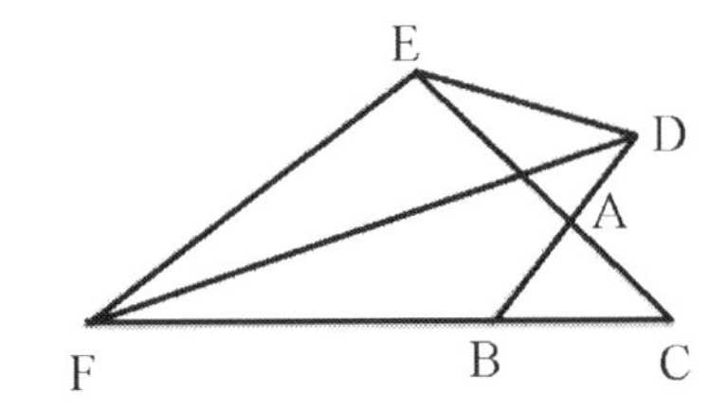

28. The difference of two positive integers is 11. The sum of the digits of one positive integer is divisible by 11. The sum of the digits of the other positive integer is divisible by 11 as well. What is the smallest possible value of the smaller integer?

29. Alex walks from village A and Bob walks from village B toward each other starting at the same time. They meet in 8 hours. If each person increases the speed by 2 km per hour, they will meet at the point 3 km away from the middle of the distance AB. What is Alex's speed if he walks faster than Bob? Express your answer as a decimal to the nearest tenth.

30. A ball of radius 10 cm is rolling along the interior wall of the container. It rolls exactly one round and back to its original position. Find the total area brushed by the ball rolling. Use 3.14 for π in your calculation.

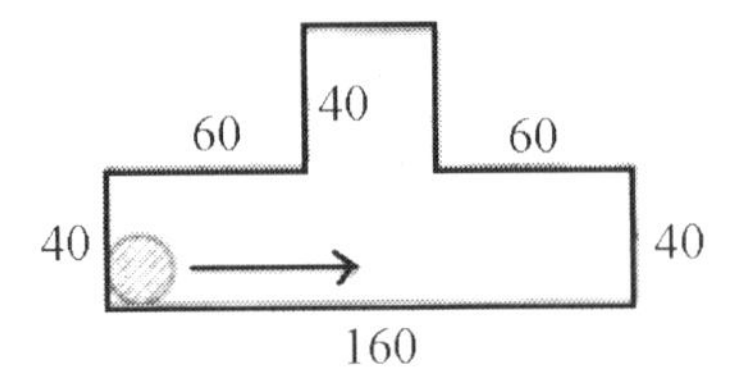

Answer keys to practice test 2

1. 37/10
2. F
3. 55
4. 81420
5. 130
6. 123
7. 36
8. 18
9. 1887
10. 18
11. 406
12. 16
13. 13
14. 24
15. 2
16. BOB
17. 550
18. 365
19. 764
20. 9/2
21. 51
22. 48 km/hour
23. 35 min.
24. 15
25. 6
26. 42
27. 7
28. 189999999999
29. 6.5 km/hour
30. 7699

MATHCOUNTS

■ Sprint Round Competition ■
Practice Test 3
Problems 1-30

Name ______________________________

Date ______________________________

DO NOT BEGIN UNTIL YOU ARE INSTRUCTED TO DO SO.

This round of the competition consists of 30 problems. You will have 40 minutes to complete the problems. You are not allowed to use calculators, books, or any other aids during this round. If you are wearing a calculator wrist watch, please give it to your proctor now. Calculations may be done on scratch paper. All answers must be complete, legible, and simplified to lowest terms. Record only final answers in the blanks in the right-hand column of the competition booklet. If you complete the problems before time is called, use the remaining time to check your answers.

Total Correct	Scorer's Initials

1. Calculate: $82 - 38 + 49 - 51$

2. MacDonald's sells a special hamburger that costs $10 each. Every Tuesday the restaurant has a discount that when you buy two hamburgers, you get one free. If Alex and his friends want to buy 9 hamburgers on Tuesday, at least how many dollars do they need to pay?

3. Bob's mom just bought 72 eggs for Bob this morning. Bob has a hen that can lay one egg per day. If Bob eats 4 eggs per day, how many days the eggs can last?

4. The sum of five counting numbers made only by the digit 8 is 1000. What is the difference between the largest number and the second largest number of the five numbers?

5. Given:

$$1 \times 9 + 2 = 11$$
$$12 \times 9 + 3 = 111$$
$$123 \times 9 + 43 = 1111$$
$$\ldots\ldots\ldots\ldots$$
$$\Delta \times 9 + \Phi = 111111$$

What is the value of $\Delta + \Phi$?=

6. What is the integer part of the expression $\frac{2010}{1000} + \frac{1219}{100} + \frac{27}{10}$?

7. Eastern Elementary School has 2400 students. Each student has five class periods every day. Each teacher has four classes to teach everyday and each classroom has exactly 30 students. Determine the number of teachers in the school.

8. Alex goes to a store to buy some pens with the money he has. He is lucky to find out that the store has a discount of 25% so he is able to buy 25 more pens with his money. How many pens could he have bought with his money originally before discount?

9. Find the shaded area in the figure with the two hearts. The large square has the side of 40 mm and the small square has the side of 20 mm. Express your answer in terms of π.

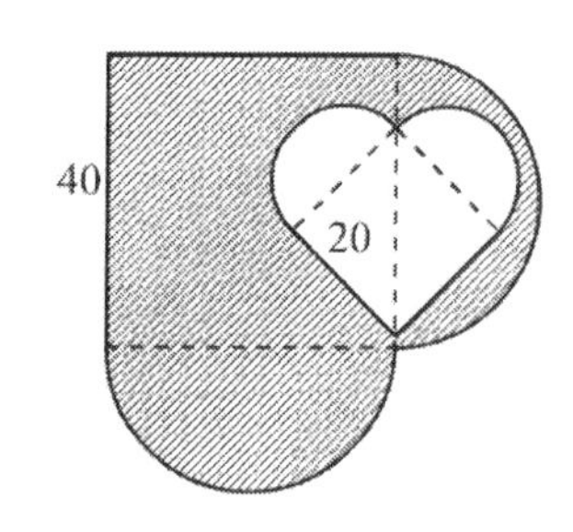

10. The product of a 2-digit integer and 4.02 is 3-digit integer. What is 10 times the 3-digit integer?

11. There are 30 days in April with five Saturdays and five Sundays. What day is the first day of April?

12. Charles throws three darts at the target with the scores in each section indicated in the figure. The score is zero if missed, What is the smallest sum of the total scores impossible for Charles to get?

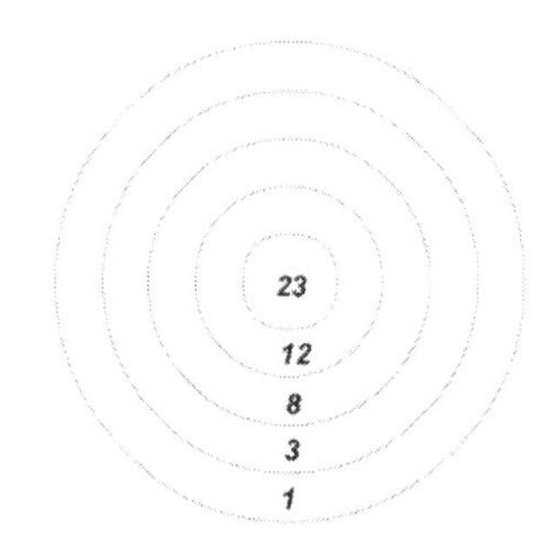

13. Two identical scales are balanced as shown in the figure by balls, triangle blocks and weighs. How many grams do the one ball and one triangle block weigh?

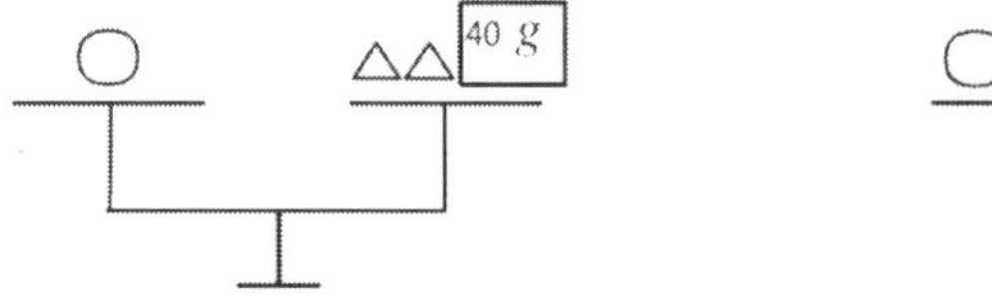

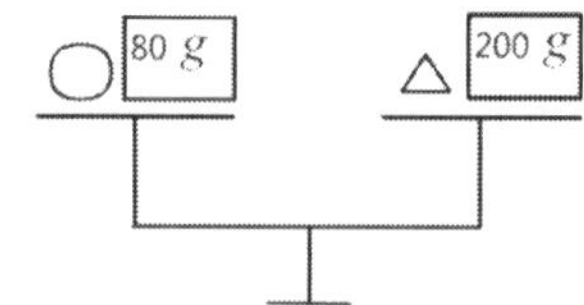

14. Find the value of $A+B+C+D+E+F+G$ if $\overline{ABCD}+\overline{EFG}=2010$. Each letter represents a different digit.

15. There are 40 boys in a group originally. Each time three boys are removed from the group while two girls are added to the group. How many times are needed so that the number of boys is the same as the number of girls in the group?

16. Bob's team right now has a winning rate of 45%. If his team can win 6 out the next 8 games, the winning rate will be increased to 50%. How many games has Bob's team already won?

17. Define $a \blacksquare b = \frac{a \times b}{a+b}$, find the value: $\underbrace{2010 \blacksquare 2010 \blacksquare 2010 \blacksquare \ldots \blacksquare 2010 \blacksquare 2010}_{\text{total } 9 \blacksquare \text{'s}}$.

18. As shown in the figure, the side of the larger square is 10 cm, and the side of the smaller square is 6 cm. Find the shaded area.

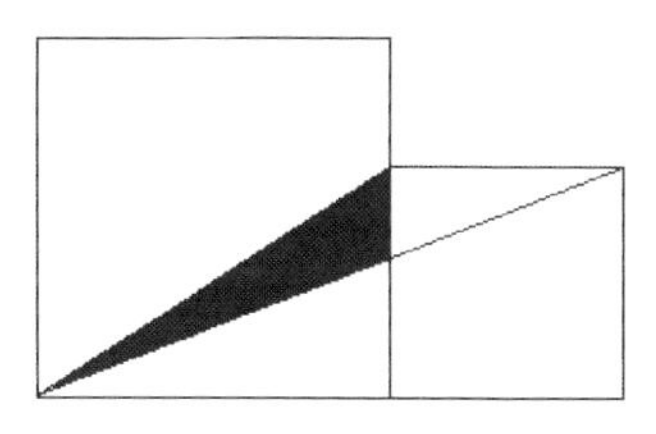

19. The sum of two positive integers is 70. Their greatest common factor is 7. What is the greatest possible value for the difference of these two positive integers?

20. There are five cards each labeled with a different number of 1, 2, 3, 4, and 5. Randomly put them on the table in a row to form a 5-digit number. What is the probability that the 5-digit number is divisible by either 5 or 2? Express your answer as a common fraction.

21. Given $a^2 + a = 1$ and $b^2 + b = 1$, where a ≠ b, find the value of $a^2b + ab^2$.

22. ΔABC with $BD = DE = EC$, $CF: AC – 1: 3$. If the areas of ΔADH is 24 square centimeters more than the area of ΔHEF, find the area of ΔABC.

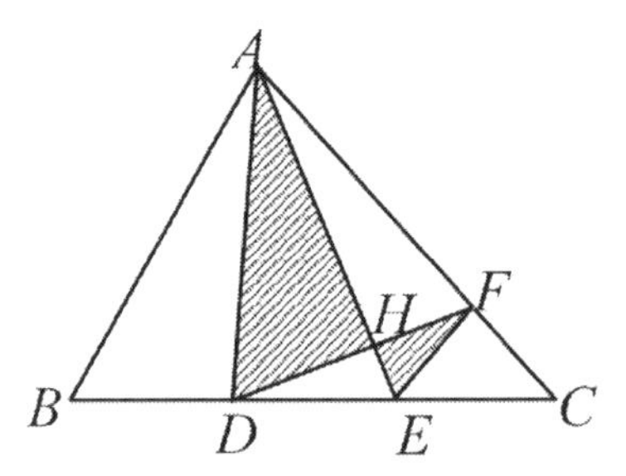

23. The positive difference between the number of factors of a positive integer and the number of factors of another integer that is two times the positive integer is 2. The positive difference between the number of factors of the positive integer and the number of factors of another integer that is three times the positive integer is 3. What is the positive integer?

24. The figure shows a 6 × 6 squares partially filled with the numbers 1 to 6. How many ways are there to fill the rest of squares with the numbers 1 to 6 such that each number appears exactly once in each row and each column.

1	2	3	4	5	6
2					5
3					4
4					3
5					2
6	5	4	3	2	1

25. Figure 1 shows a 3 by 3 squares. Each small square contains an arrow and a number. The direction of the arrow points to another square where the next consecutive number is located. Note that these two squares are not necessarily next to each other. A cycle is finished if you complete the move from the square marked "1" to the square marked "9". Figure 2 also shows a cycle of the move with some numbers missing. Find the number in the small square marked with the letter "A". Note that the symbol "★" means the cycle is finished.

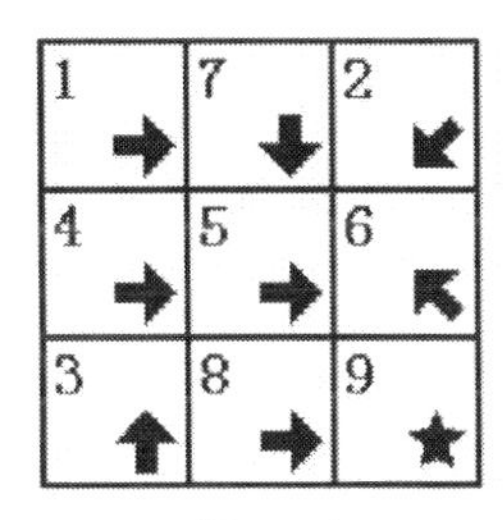

Figure 1

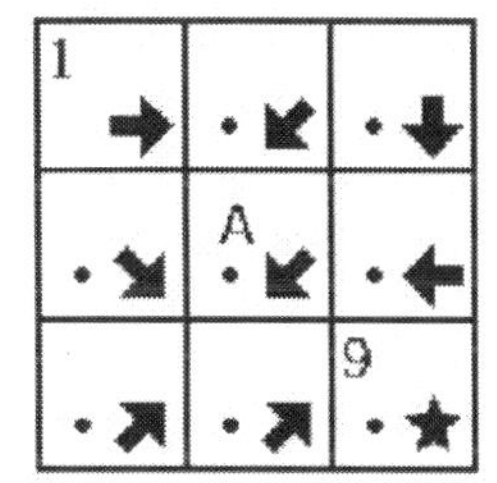

Figure 2

26. Figure 2 (8×5 rectangle) is obtained by turning Figure 1 180°. If figure 2 is put on the top of figure 1 such that they are completely overlapped, find the number of 1×1 shaded squares that are overlapped.

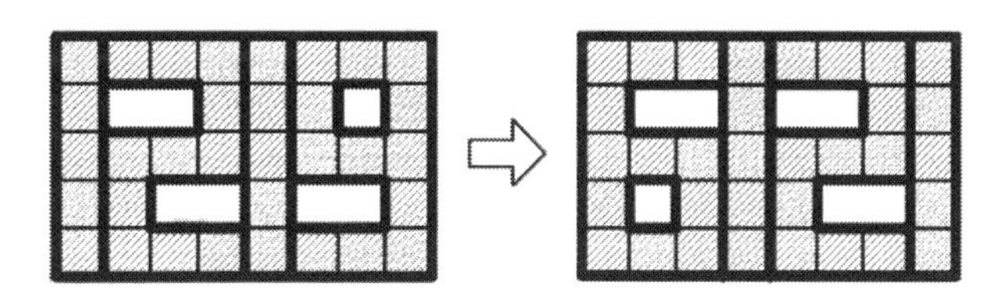

Figure 1 Figure 2

27. Alex, Bob, Charles, and Danny finished a math test consisting of 10 questions. Each person got exactly 8 questions correct. At least how many questions did they all get correct?

28. 2010 students line up in a column. Each student takes turns from number one to number 2010 to say a number following the rule below: if the number a student says is an one-digit number, the next student must say another number that is the sum of the number and 8; if the number a student says is a two-digit number, the next student must say another number that is the sum of the units digit of the number and 7. Now the number the first student says is 1, what is the number that the last student says?

29. There are three flowers in a greenhouse.

(1) There is only one day in a week that three flowers will bloom together.

(2) No flower can bloom for three consecutive days.

(3) It will not exceed one day in a week such that any two flowers do not bloom at the same time.

(4) Flower A will not bloom on Tuesday, Thursday, and Sunday.

(5) Flower B will not bloom on Thursday and Saturday.

(6) Flower C will not bloom on Sunday.

What day do the three flowers bloom at the same time?

30. A cylinder's height is three times its radius. The cylinder is cut into two parts as shown. If the surface area of the resulting larger cylinder is 3 times the surface area of the resulting smaller cylinder, what is the ratio of their volumes?

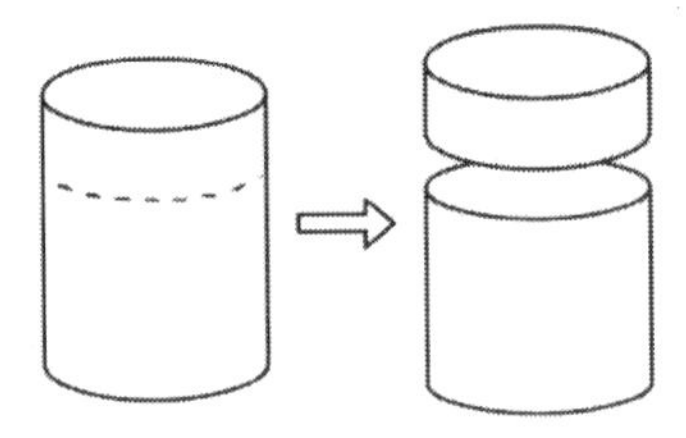

Answer keys to practice test 3

1. 42
2. 60
3. 24
4. 800
5. 12351
6. 16
7. 100
8. 75
9. $1200 + 300\pi$ mm^2
10. 2010
11. Saturday
12. 22
13. 280
14. 30
15. 8
16. 18 cm^2
17. 201
18. 45/4
19. 56 (there are two possible values for the difference: 56 and 28).
20. 3/5 ($P = \dfrac{4! + \binom{2}{2}4!}{5!} = \dfrac{3}{5}$)
21. 1
22. 108 cm^2
23. 12
24. 16
25. 6
26. 30
27. 2
28. 13
29. Friday
30. 11

MATHCOUNTS

■ Sprint Round Competition ■
Practice Test 4
Problems 1-30

Name

Date

DO NOT BEGIN UNTIL YOU ARE INSTRUCTED TO DO SO.

This round of the competition consists of 30 problems. You will have 40 minutes to complete the problems. You are not allowed to use calculators, books, or any other aids during this round. If you are wearing a calculator wrist watch, please give it to your proctor now. Calculations may be done on scratch paper. All answers must be complete, legible, and simplified to lowest terms. Record only final answers in the blanks in the right-hand column of the competition booklet. If you complete the problems before time is called, use the remaining time to check your answers.

Total Correct	Scorer's Initials

1. Calculate: $(11 + 12 + 13 + 14 + 15 + 16 + 17) \div 7$

2. If the pattern continues: $2\Delta3 = 2 + 3 + 4$, $5\Delta4 = 5 + 6 + 7 + 8$,…., what is $3\Delta5$?

3. Alex has fourteen $5 and $10 bills totaling $100. How many of these bills are $5?

4. Find the sum of all positive two-digit odd integers not divisible by 9.

5. There are five Mondays and four Tuesdays in August of a year. What day is August 8th of the year?

6. Select three numbers from 2, 5, 7, and 9 to form a 3-digit number that has the remainder 2 when divided by 3. How many such 3-digit numbers are there?

7. What is the ratio of the rotational speed of the second hand to that of the hour hand of a clock?

8. Alex, Bob, and Charlie are students who participate in three different clubs (Math, Science, and Writing) from three schools (Hope School, Cox School, and Ashley School). Alex is not from Hope School, Bob is not from Cox School. The student from Hope School is not in the Science club, the student from Cox School is in the Math club. Bob is not in the Writing club. Which school is Charlie from and what club is he in?

9. Twenty five apples are distributed to n kids such that no matter how the apples are distributed, it is always true that at least one kid gets 7 apples or more. What is the value for n?

10. There are 40 students in a class. 32 of them took the math exam in the morning and 24 of them took the reading exam in the afternoon. There are 20 students who took both exams. How many students did not take any exam?

11. As shown in the figures, there are 6 small circles in the first figure, 10, 16, and 24 circles in the successive figures. If the pattern follows, how many small circles are in the sixth figure?

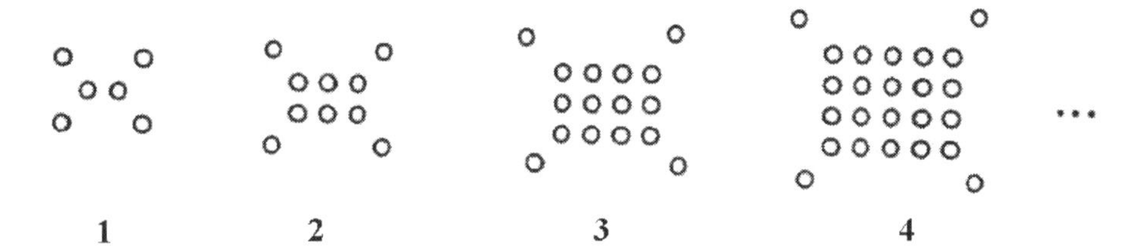

12. In a map with the ratio of 1:50,000,000, the distance is 2 cm between Apex and Burham, and 3 cm between Apex and Cali. Find the greatest possible real distance in kilometers between Burham and Cali.

13. ABCD is a square. The diagonals meet at point O. Draw isosceles triangles using the sides AD, AB, and BC as shown in the figure. The number of right triangles is m and the number of squares is n. What is $m + n$?

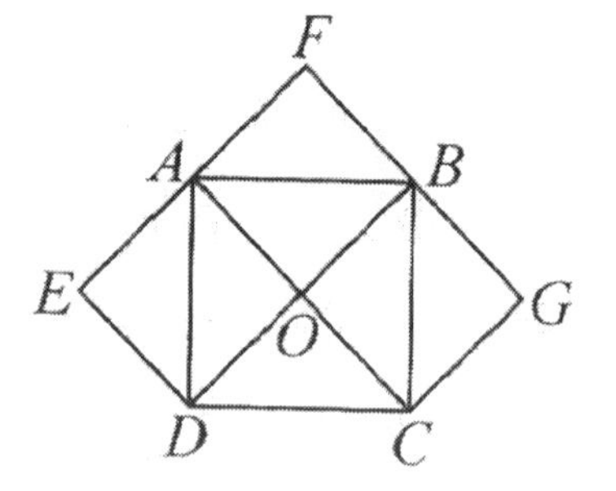

14. There are two rabbits in a racing tournament. One is big and one is small. The small rabbit runs 10 steps before the big rabbit starts to run. The time used by the small rabbit to run 4 steps is exactly the same as the time used by the big rabbit to run 3 steps. The distance covered by the small rabbit in 7 steps is exactly the same as the distance covered by the big rabbit in 5 steps. How many steps does the big rabbit need to run in order to catch the small rabbit?

15. The store's rule is that you can use 3 empty cans to exchange for 1 can of soda. Alex's class has 17 kids and each bought a can of soda at the beginning of the party. At most, how many more cans of soda can Alex's class get by exchanging?

16. A wall with a hole is shown in the figure. How many bricks are needed to repair it?

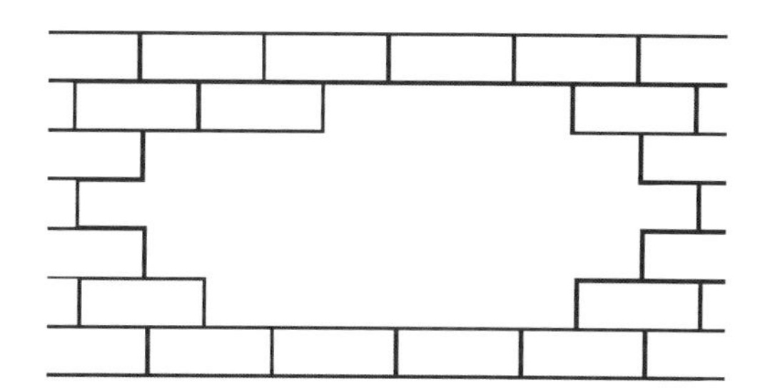

17. There are three small circles in a big circle as shown in the figure. Fill each small circle with one of the numbers 1 to 7 such that n, the sum of the three numbers in each line is the same as the sum of three numbers in each big circle. Find n. (Each number can only be used once).

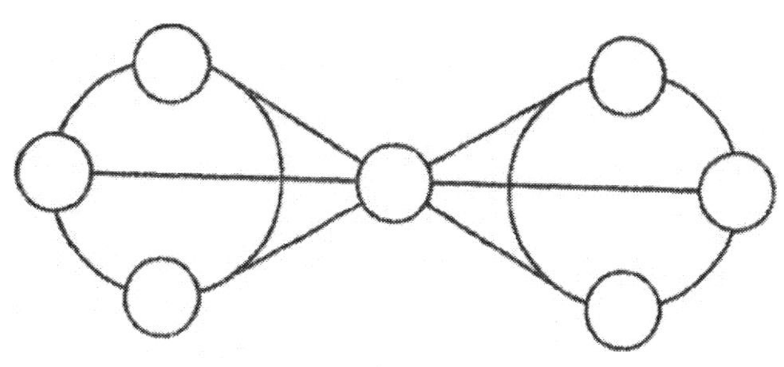

18. m is the smallest positive integer and n is the greatest positive integer such that the sum of the digits of each positive integer is 15 and each digit is different. What is $m + n$?

19. An escalator moves up at a constant speed connecting the first floor and the second floor. Two children walk up on the escalator. The boy walks 20 stairs in 1 minute, and the girl walks 15 stairs in a minute. The boy reaches the second floor in 5 minutes and the girl reaches the second floor in 6 minutes. How many stairs are visible when the escalator is still?

20. A TV company needs to assemble 2400 TV sets. Two groups start to work at the same time. Group A can assemble 60 sets each day and Group B can assemble 62 sets each day. How many days have passed when there are still 204 sets left unassembled?

21. At most, how many regions can 6 lines divide a rectangle?

22. There are 3 consecutive odd positive integers. The positive difference between the product of the first two odd integers and the product of the last two odd integers is 252. What is the smallest number of the three integers?

23. A number of identical unit cubes is used to form a solid. Figures below show the different views of the solid. The first figure is the front view, second is the left view and the third one is the top view. How many cubes are there?

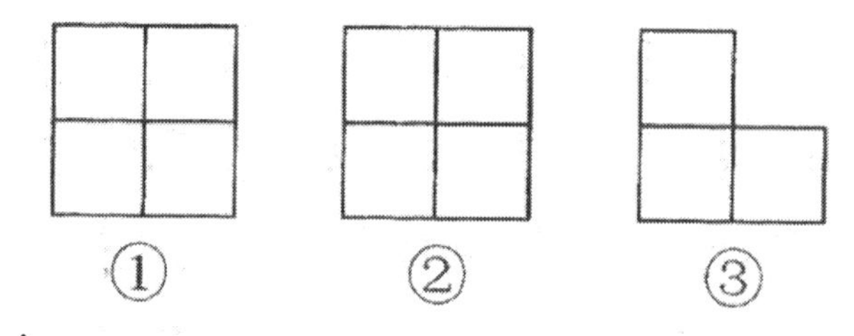

24. A square shape garden is shown in the figure. Find the area of the shaded portion.

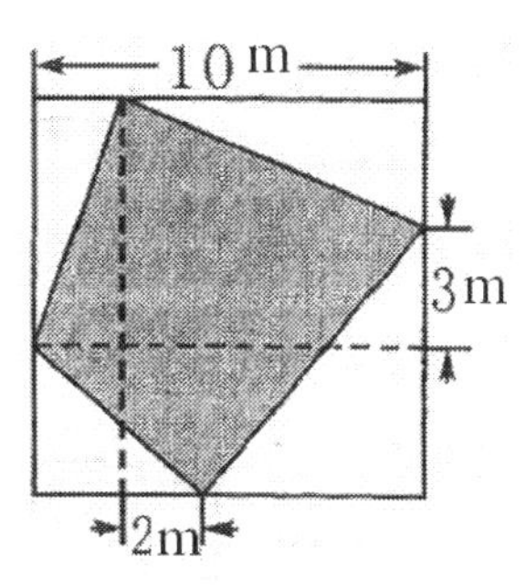

25. A tank is filled with some water. If the weight of the water is doubled, the total weight is then 10 kg. If the weight of the water is five times the weight of the water originally, the total weight is then 19 kg. Find the weight of the water originally in the tank.

26. There are some positive integers m and n such that the product of $(m-n)$ and $(m+n)$ is 77. What are these positive integers?

27. Mouse A and Mouse B are separated by a distance of 1.62 meters underground. They decide to meet by digging all the way through. Mouse A will double his speed every day, that is, he starts to dig 2 cm the first day, 4 cm the second day, and so on. Mouse B will dig at a constant speed, that is, 6 cm per day. How many centimeters will Mouse A have dug when they finally meet?

28. Eastern Middle School has three 8th grade classes: A, B, and C. The school has a competition for math, reading and science for all 8th graders. Each subject gives awards to the top three students: first place 5 points, second place 3 points, and third place 1 point. It is known that the number of students in class A getting into the top three is the smallest. The number of students in class B getting into the top three is twice as many as the number of students in class A getting into top three. But, the total number of points for these two classes is the same and they both ranked first (tie). How many points does class C get?

29. Alex, Bob, and Charlie together have \$168. They start to exchange their money in the following way: at first, Alex gives Bob some money that is exactly the same as the amount Bob has. Bob then gives Charlie some money that is exactly the same as the amount Charlie has. In the end, Charlie gives some money that is exactly the same as the amount Alex has at that time. After the exchanges, each person has the same amount of money as the others. How many more dolloars does Alex have at the beginning than Bob?

30. Train A 's speed is 20 meters per second and Train B 's speed is 14 meters per second. If they start at the same time with their heads at the same starting points but on different rails, train A can passes train B completely in 40 seconds. If train A and train B start at the same time with their tails at the same starting points, train A can pass train B completely in 30 seconds. Train A's length is m and train B 's length is n. What is the value $m + n$?

Answer Keys to practice test 4

1. 14
2. 25
3. 8
4. 2160
5. Saturday
6. 6
7. 720:1
8. Hope School, Writing.
9. 4
10. 4
11. 46
12. 2500 km
13. 11+4=15
14. 150
15. 8. (5+2+1)
16. 18
17. 12
18. 69+543210=543279
19. 150
20. 18 days
21. 22
22. 61
23. 6 or 5, both answers are correct.
24. 53 m^2
25. 3 kg
26. (2, 9) and (38, 39)
27. 126 cm
28. 7 points
29. $28
30. 240+180=420 m

MATHCOUNTS

■ Sprint Round Competition ■
Practice Test 5
Problems 1-30

Name

Date

DO NOT BEGIN UNTIL YOU ARE INSTRUCTED TO DO SO.

This round of the competition consists of 30 problems. You will have 40 minutes to complete the problems. You are not allowed to use calculators, books, or any other aids during this round. If you are wearing a calculator wrist watch, please give it to your proctor now. Calculations may be done on scratch paper. All answers must be complete, legible, and simplified to lowest terms. Record only final answers in the blanks in the right-hand column of the competition booklet. If you complete the problems before time is called, use the remaining time to check your answers.

Total Correct	Scorer's Initials

1. Calculate: $100-98+96-94+92-90+\cdots\cdots+4-2$

2. If the pattern continues, what is the value of x?

2	3
4	5

1

5	7
9	11

2

8	11
14	x

3

3. Twenty pounds of apples and thirty pounds of pears cost $132. Two pounds of apples costs the same as 2.5 pounds of pears. How much does one pound of apples cost?

4. There is only one person who knows French among Alex, Bob, and Charlie.

Alex: I know French

Bob: I do not know French.

Charlie: Alex does not know French.

If there is only one person who speaks the truth, which person is it?

5. The average value of 10 numbers is 789. The average of eight of those ten is 678. Find the average of the other two numbers.

6. If the difference of two 4-digit positive integers is 3456, it is said that they form a pair. How many such pairs are there?

7. Fill out four blank regions as shown in the figure with one of the digits: 3, 5, 7, and 8 such that the sum of the four numbers in each circle is 21. Which number needs to be in the middle region?

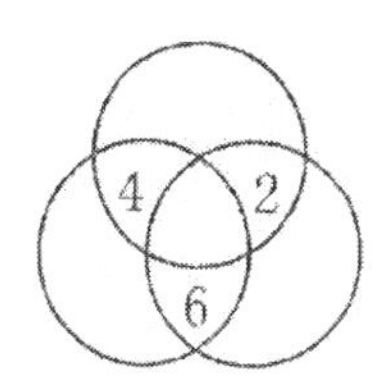

8. Five pieces of square papers of side length of 10 cm are stacked such that the overlapping part of the two squares is one fourth of the area of a square. Find the perimeter of the resulting figure.

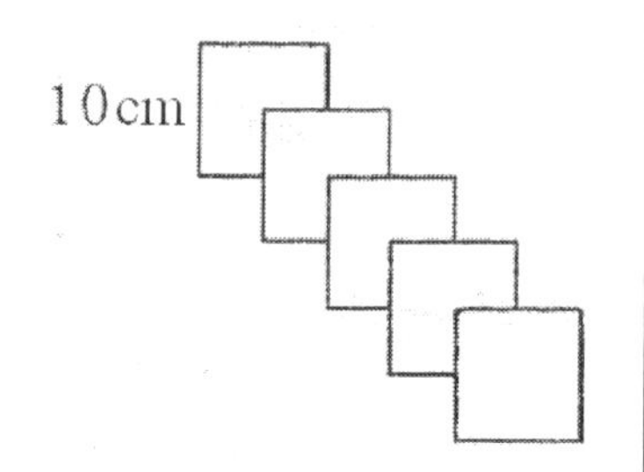

9. Five teams A, B, C, D and E are having a soccer tournament with each team playing once with every other team. Up to today, team A has played 4 games, team B has played 3 games, team C has played 2 games, and team D has played one game. How many games has team E played?

10. Shown in the figure is a plot of scores in three subjects (Music, Sport, and Art) for three teams A, B, and C. Which team should be ranked first based on total points earn?

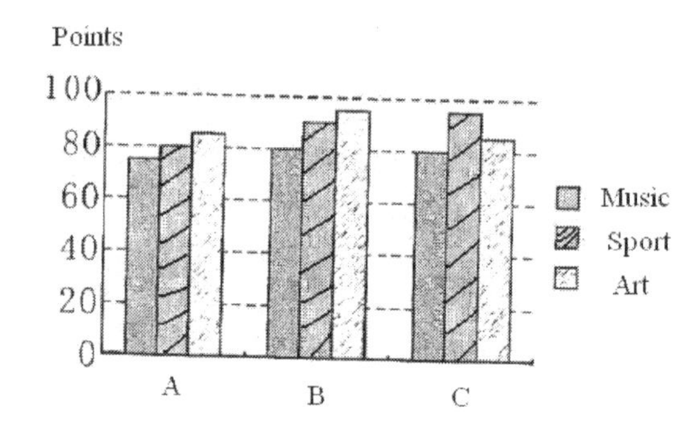

11. Ayden and Kinston are 720 meters apart. Alex walks from Ayden toward Kinston, and Bob walks from Kinston toward Ayden, at the same time. After 8 minutes they are 200 meters apart. Alex's speed is 30 meters per minute. What is the positive difference between Bob's speed and Alex's speed?

12. The figure shows the numbers of days needed for each person (Alex, Bob, and Charlie) to make 3,000 toys. They need to make n toys. The three people work together for 5 days, then Alex and Bob work together for three more days to finish the job. Find n.

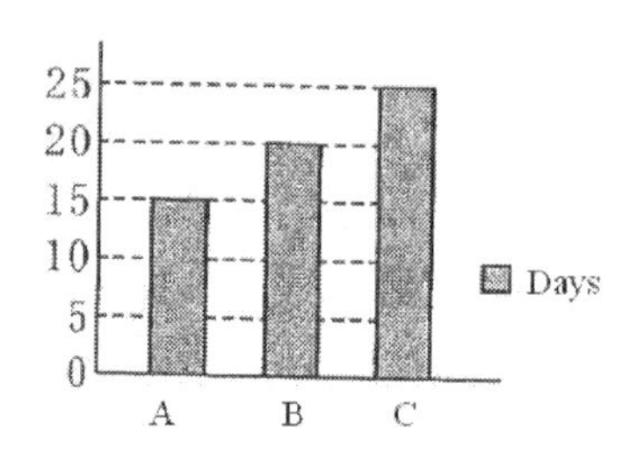

13. Ninety six students are arranged in a rectangular shape. There are 6 students in each row. The distance is 1 meter between every two students. What is the area of the rectangle?

14. As shown in the figure, seven congruent small circles are fit exactly in the large circle with a radius of 3. All the circles are tangent to one or another. Find the shaded area.

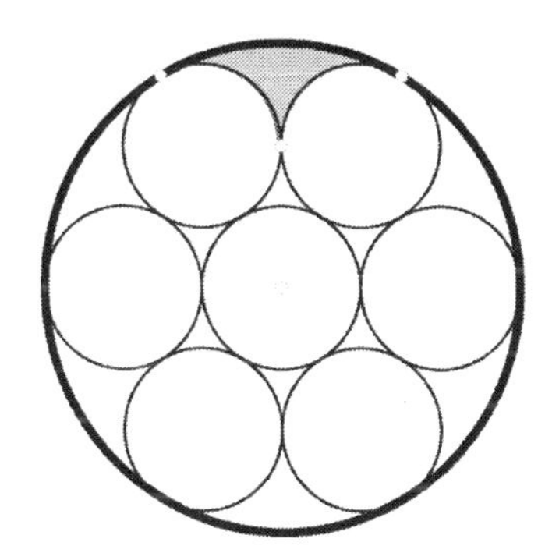

15. Alex puts 127 marbles into n bags in a way such that no matter how many marbles (1 to 127) you ask for, he is always able to meet your requirement immediately by handing you one or more bags without taking out any marble from any bag. What is the smallest value for n?

16. There are 48 students in Bob's class. 23 of them participate in Science Club, 26 participate in Art club, and 12 participate in both clubs. How many students do not participate in any of the two clubs?

17. The sum of the ages of Bob, his's father, mother, and older sister is 83 this year. The sum of their ages 5 years ago was 65. How old is Bob now?

18. Place each of the digits 2, 3, 6, 8, and 9 in one of the boxes to make a two-digit number and a 3-digit number such that the product is the largest you can produce.

19. How many ways are there to use four numbers from 1 to 10 such that the sum of the four numbers is odd?

20. Arrange the positive integers 1, 2, 3, 4,… as shown in the figure. What is the number in the first row of the 20th column?

1	3	6	10	15
2	5	9	14	20
4	8	13	19	
7	12	18		
11	17			
16				

21. A, B, C, and D are four positive integers in order from smallest to the greatest. Taking two of them at a time and adding the pair together will result in 5 different numbers: 21, 23, 24, 25, and 27. What is the average value of A, B, C, and D?

22. A rectangular solid placed on the ground with one side attached to a wall is formed by gluing unit cubes together. The rectangular solid has the length 40, width 20, and height 16. If the rectangular solid is painted red on all visible faces and then dissembled, at most how many unit cubes have no faces painted red at all?

23. There are four regions (each region has four teams) in a soccer tournament. The selection rule is that each team plays each other team within the region once. The top team will represent the region in the final tournament. Each of the four teams in the final competition plays each other team twice. How many games are there total?

24. A road 50 meters long is placed with flowers on both sides of the road. The distance between any two neighboring flowers is 2 meters. At most, how many flowers can be placed?

25. Find the sum of the digits in the quotient when $\underbrace{1234112341\ldots\ldots12341}_{\text{repeated 20 times}}$ is divided by 7.

26. Two identical white square bricks and two identical black rectangle bricks are placed as shown in the figure. Find the area of the white square brick.

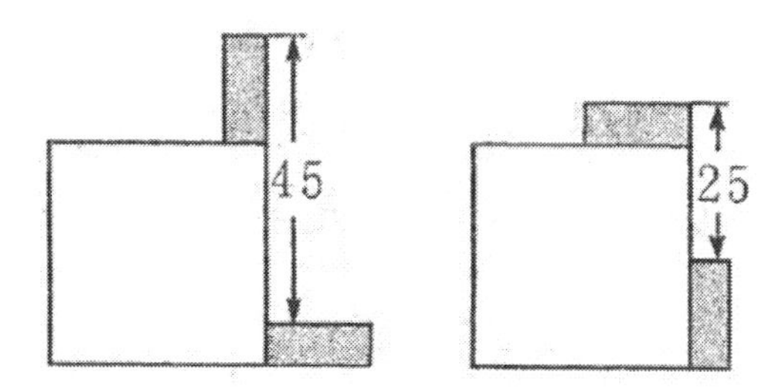

27. Candy is placed inside one of the five boxes labeled 1 to 5. Each box has attached a note as follows:

Box 1: Candy is in box 5.

Box 2: Candy is inside box 4 or 5.

Box 3: Candy is not in box 3.

Box 4: Candy is not inside the box 5.

Box 5: Candy is inside box 4.

Alex is told that only one note is correct. Which box has the candy?

28. There are 200 lamps labeled 1 to 200 and each is controlled by an independent switch which shows "off" position right now. There are 200 kids and each is numbered 1 to 200. Each kid needs to operate every switch once that is a multiple of his/her number. How many lamps are on after every kid finishes?

29. Use the digits a, b, and c without repetition to form six 3-digit positive integers. The sum of these six positive integers is 3774. It is known that the greatest digit of a, b, and c is twice as much as the smallest one. What is the greatest of the six positive integers?

30. Tommy Turtle and Robby Rabbit are having the big race of 1000 meters. Robby can run 40 meters in one minute and will rest 25 minutes for every 5 minutes running. Tommy can run 8 meters in one minute and will not rest at all. What will be result of the race?

Answer Keys to practice test 5:

1. 50
2. 17
3. $3/lb
4. Charlie
5. 1233
6.5544
7. 8
8. 120 cm
9. 2

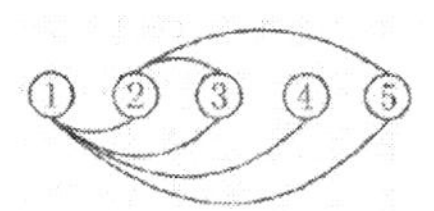

10. B
11. 5 meters
12. 3400
13. 75
14.

$\frac{5}{6}\pi - \sqrt{3}$

15. 7

$127 = 1 + 2 + 4 + 8 + 16 + 32 + 64$

16. 11. (48 – 37)
17. 3 years old
18.

93×862 or $80{,}166$

19. 100, $\binom{5}{1} \times \binom{5}{3} + \binom{5}{3}\binom{5}{1}$.

20. 210
21. 12
22. 10830 (10530 and 10830)
23. 36 (4×6+2×6)
24. 52
25. 340
26. 1225 cm^2
27. 3
28. 14
29. 854
30. tie

MATHCOUNTS

■ Sprint Round Competition ■
Practice Test 6
Problems 1-30

Name

Date

DO NOT BEGIN UNTIL YOU ARE INSTRUCTED TO DO SO.

This round of the competition consists of 30 problems. You will have 40 minutes to complete the problems. You are not allowed to use calculators, books, or any other aids during this round. If you are wearing a calculator wrist watch, please give it to your proctor now. Calculations may be done on scratch paper. All answers must be complete, legible, and simplified to lowest terms. Record only final answers in the blanks in the right-hand column of the competition booklet. If you complete the problems before time is called, use the remaining time to check your answers.

Total Correct	Scorer's Initials

1. Calculate $\frac{7}{12} \div \left[\frac{1}{4} \div \left(\frac{1}{3} - \frac{1}{4}\right)\right] \times 12$. Express your answer as a common fraction.

2. A new fraction is formed when the numerator of a fraction is reduced 25% while the denominator is increased 25%. What is the percent of reduction of the new fraction to the original fraction?

3. When the fraction $\frac{2}{7}$ is converted into the decimal form, what is the 2011th digit after the decimal point?

4. If $\underbrace{20112011\ldots\ldots2011}_{\text{repeated } n \text{ times}}20$ is divisible by 18, find the smallest value for *n*.

5. If $35p + 13q = 135$, where both p and q are prime numbers, find the values of p and q.

6. $A * B = A \times 2 - B \times 3 + A \times B$. Find $5*3$.

7. The sum of the numerator and denominator of a fraction in the simplest form is 86. If both the numerator and the denominator are 9 less, the new fraction is $\frac{8}{9}$. What is the original fraction?

8. How many 3-digit positive integers with three distinct digits are divisible by 15?

9. A good number has the property that when it is divided by 13, the quotient and the remainder are the same. How many good numbers are there between 1 and 200?

10. Four nets are labeled (1), (2), (3), and (4) as shown in the figure. Which one can be folded to form the cube?

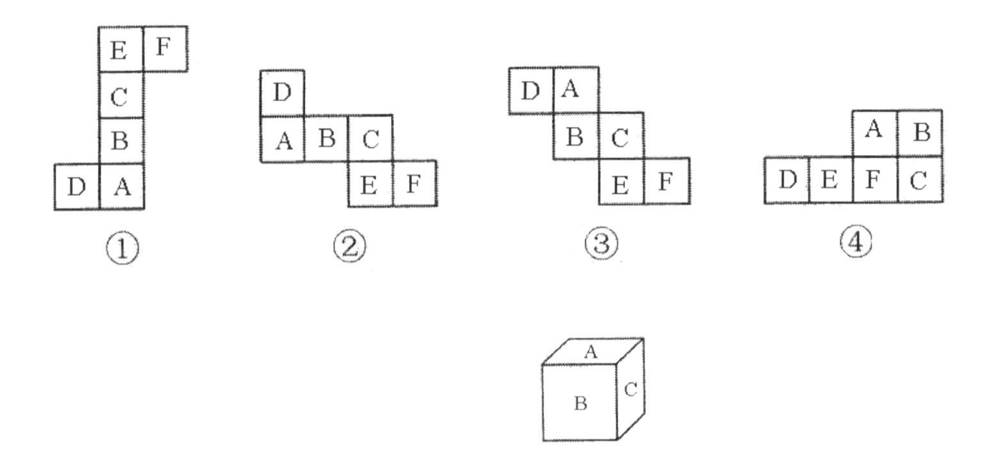

11. As shown in the figure, both ABCD and DEFG are squares. ED = 2AB. The area of the triangle BEG is 6. What is the area of the square ABCD?

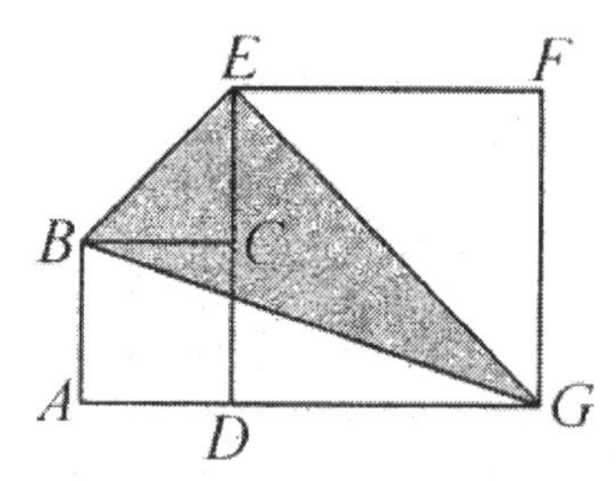

12. The perimeter of a rectangle is 48 cm. The length and width of the rectangle are in the ratio 5:3. Find the area of the rectangle.

13. ABCD is a square. AB = 12 cm. AE = 5 cm. DEFG is a rectangle. FG = 13 cm. Find DG. Express your answer as a common fraction.

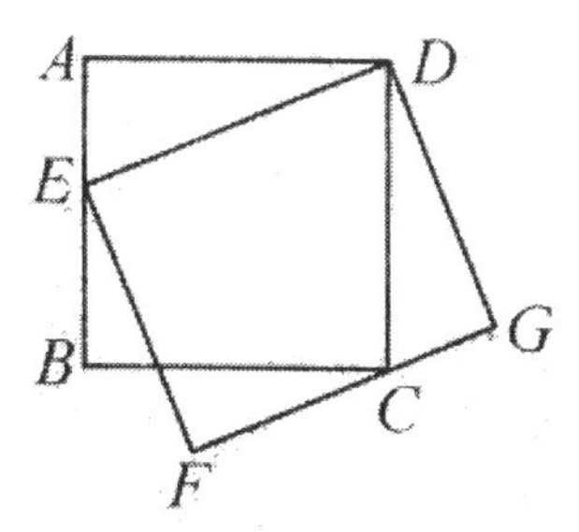

14. As shown in the figure, a sector with the center angle of 60° is cut off the circle. Find the ratio of the areas of the shaded portion to the sector that is cut off.

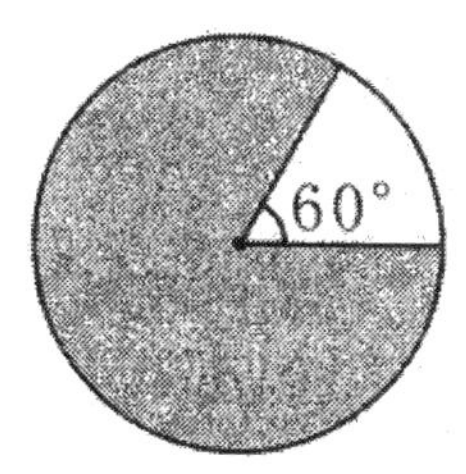

15. A right circular cone-shaped cup contains 3 liters of water. The water level is exactly half of the height of the cup. How many more liters of water are needed to make the cup full of water?

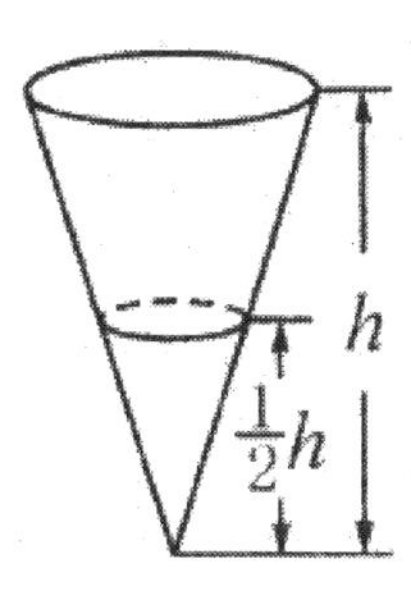

16. The sum of the edges of a rectangular prism is 24 cm. The three dimensions of the rectangular prism are distinct integers. Find the volume of the prism.

17. The time right now is 8:30 in the morning. What will be the time after the second hand turns 2011 revolutions?

18. Alex alone can finish a job in 5 hours. If Alex and Bob work together, the job can be done in 2 hours. How many hours can Bob finish the job alone? Express your answer as a common fraction.

19. A rectangular prism with two square bases is cut into two congruent cubes. The sum of the surface areas of the two cubes is 98 cm^2 more than the surface area of the rectangular prism. Find the volume of the rectangular prism.

20. The price of shirt A is increased by \$0.16 which is 16% of the original price. The price of shirt B is increased by \$3.36 which is also 16% of the original price. What is the positive difference of the two original prices for these two shirts?

21. Alex, Bob, and Catherine evenly divide a bag of candy. Each person eats 8 pieces of candy. The sum of the number of candy left is exactly the number of candy each person gets at the beginning. How many pieces of candy are there in the bag originally?

22. A water container is full of water now. 10% of water is used in the morning. 59 liters are used in the afternoon. 20% of water remaining is used in the evening. One liter less than half of water in a full container remains in the end. What is the volume of water in the beginning?

23. Alex bought 99 adult and children tickets with \$280. Children tickets cost \$2 each and adult tickets cost \$3 each. How many children tickets are bought?

24. A sixth grade class has 36 students. The number of boys is $\frac{7}{12}$ of the number of total students. 20% of the girls are 12 years or older. How many girls are less than 12 years old?

25. 2011 boxes containing marbles are arranged in a row from left to right. The first box contains 9 marbles. The third box contains 6 marbles. Every four neighboring boxes have a sum of 32 marbles. How many marbles are there in the last box?

26. How many paths are there from A to E if no point or line segment can be traced twice?

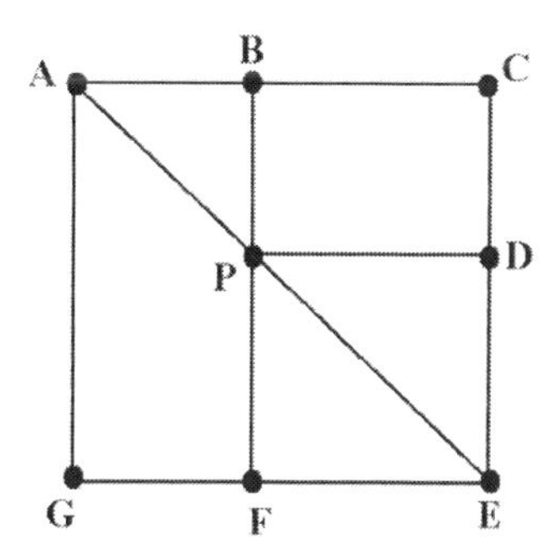

27. There are nine distinct regions within the 5 rings as shown in the picture below. Peter is going to place a different number from 2 to 10 in each of the nine regions so that the sum of the numbers inside any ring is 15. Two numbers are already placed as shown. What number should be D?

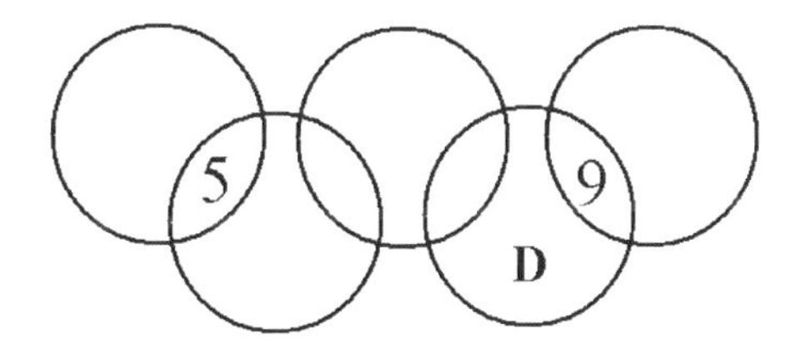

28. Bob needs to paint the exterior surface of a half cylinder log lying on the ground. What is the total area he needs to paint if he does not want to move the log? Express your answer in terms of π.

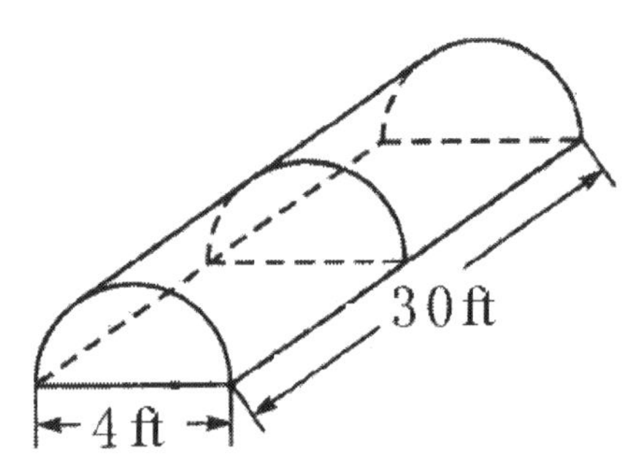

29. A marching band of 800 meters long moves at a constant speed of 80 meters per minute. Alex runs from the front to the end of the marching band with a constant speed that is 3 times the speed of the marching band. He talks to the last person in the end of the marching band for one minute and runs back to the front with the same speed. Find the total minutes needed for Alex to run both ways. Express your answer as a decimal to the nearest tenth.

30. Seven lamps are arranged in a row as shown in the figure below. Each lamp has its own switch. Now lamps A, C, E, and G are on and other lamps are off. Ben starts to turn each switch from A to G the following way: if the lamp is on, he turns it off; if the lamp is off, he turns it on. He repeats the pattern until he turns the switches 2011 times. Which lamps are on finally?

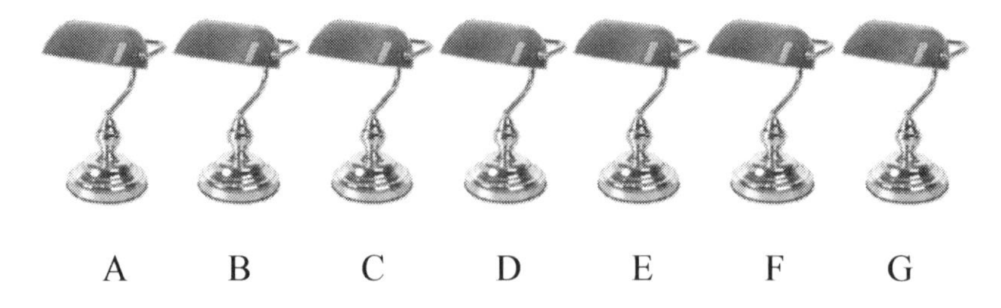

A B C D E F G

Answer keys to practice test 6

1. 7/3
2. 40%
3. 2 (2/7 = 0.<u>285714.</u> 2011 ÷ 6 =335 R1)
4. 4
5. p = 2, q = 5
6. 16

7 41/45

8. 47. For a 3-digit positive integer abc, when c = 5, we have 23 such integers; when c = 0, we have 24 of these integers.
9. 12

10.3.

11.3 cm^2

12. 135
13. 144/13
14. 5:1
15. 21
16. 6 cm^3
17. 6:01 pm
18. 10/3 hours
19. 686 cm^3
20. $20
21. 36
22. 210 liters
23. 17
24. 12
25. 6
26. 14
27. 2

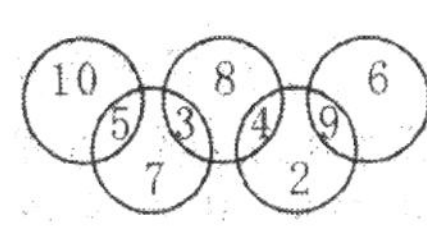

28. 64π
29. 8.5 min.
30. A, D, F (2011 = 7×287+2, AB no change, CDEFG change).

MATHCOUNTS

■ Sprint Round Competition ■
Practice Test 7
Problems 1-30

Name

Date

DO NOT BEGIN UNTIL YOU ARE INSTRUCTED TO DO SO.

This round of the competition consists of 30 problems. You will have 40 minutes to complete the problems. You are not allowed to use calculators, books, or any other aids during this round. If you are wearing a calculator wrist watch, please give it to your proctor now. Calculations may be done on scratch paper. All answers must be complete, legible, and simplified to lowest terms. Record only final answers in the blanks in the right-hand column of the competition booklet. If you complete the problems before time is called, use the remaining time to check your answers.

Total Correct	Scorer's Initials

1. Calculate $\frac{2}{5}\times\frac{1}{4}+\frac{1}{2}\div\frac{5}{8}+3.2\times\frac{1}{32}$

2. a is a digit from 1 to 9. How many times a is $a+\overline{aa}+\overline{aaa}+\ldots+\overline{aaaaaa}$?

$\overline{aa}$ means a 2-digit positive integer.

3. M is a positive integer greater than 10. $M+2$ is a multiple of 3; $M+3$ is a multiple of 4; $M+4$ is a multiple of 5; and $M+5$ is a multiple of 6. What is the smallest M?

4. When A is divided by 2011, the remainder is B and the quotient is 2010. What are the last two digits of A if B gets the greatest possible value?

5. There are n numbers in a group and the sum of any two numbers in this group is divisible by 46. These numbers are selected from 1 to 2011. What is the greatest possible value for n?

6. Positive integers 1 to 2011 are listed in the table below. The number 2011 is in the mth row and nth column. Find the value of $m+n$.

1	2	3	4	5	6	7
	13	12	11	10	9	8
14	15	16	17	18	19	20
	26	25	24	23	22	21
27	28	29	30	31	32	33
…	…	…	…	…	…	…

7. $A\div B\div C=6$. $A\div B-C=15$. $A-B=17$. Find $A\times B\times C$.

8. What is the 100th term in the list?

$$\frac{1}{2}, \frac{1}{4}, \frac{3}{4}, \frac{1}{6}, \frac{3}{6}, \frac{5}{6}, \frac{1}{8}, \frac{3}{8}, \frac{5}{8}, \frac{7}{8}, \ldots\ldots$$

9. In a country there are coins valued at 1¢, 2¢, 5¢, and 10¢. You are a visitor of that country and you have 4 coins with you (one 1¢, one 2¢, one 5¢, and one 10¢). How many different payments can you make using these 4 coins?

10. The product of two positive integers is 735. The product of one of the two integers and 3 less than the other integer is 672. Find the two integers.

11. ABCD, AEFG, GHIJ, and JKLM are squares. Points E, H, and K are the midpoints of AD, GF, and JI, respectively. AB = 8 cm. Find the perimeter of the shape formed by BCDEFHIKLM.

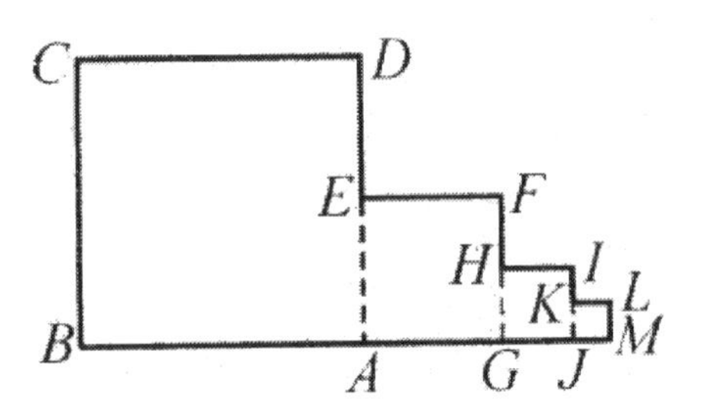

12. Find the area of the circle if the shaded area is 5 square units. Express your answer in terms of π.

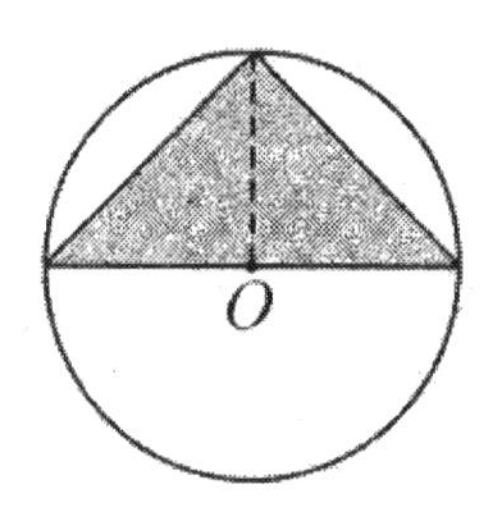

13. An isosceles trapezoid is formed by cutting off the top part of an isosceles right triangle with the long base of 10 cm and the short base of 4 cm. Find the area of the trapezoid.

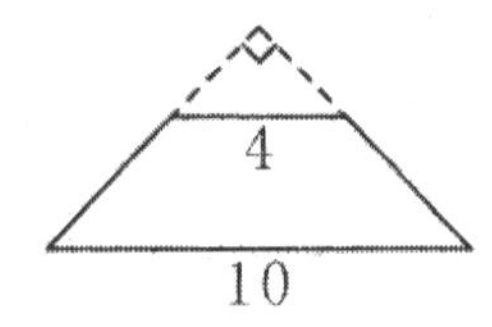

14. The length is 2 times the width of a rectangle. The diagonal is 9 cm. Two of the rectangles are put together as shown in the figure. What is the total area? Express your answer as a decimal to the nearest tenth.

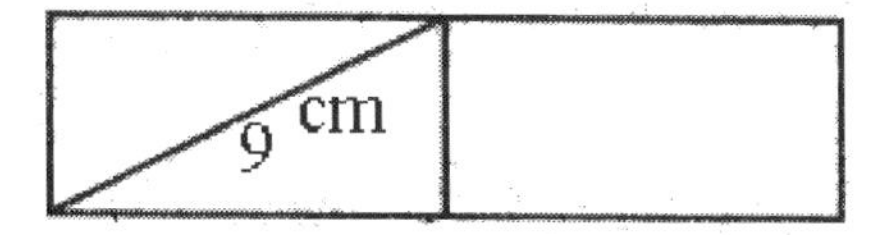

15. Find the perimeter of the shaded area as shown in the figure. Express your answer in terms of π.

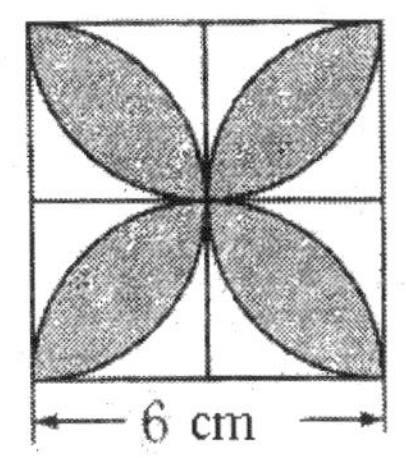

16. A rectangular prism is formed by gluing together 9 unit cubes. Ignoring the units, what is the greatest possible sum of the surface area and the volume of the rectangular prism?

17. A radio is originally priced at $50. This price is then reduced twice and becomes $40.50. If the percent of two reductions is the same, what is it?

18. Alex finished $\frac{1}{5}$ of his homework problems in the first week and 30% in the second week. It is known that the number of problems solved in the second week is 30 more than the problems solved in the first week. How many problems did Alex have for his homework?

19. Alex adds salt to a 25% salt solution of 1000 grams and the concentration of the solution becomes 50%. Later he wants to get a 30% solution. How many grams of water does he need to add?

20. There are 40 students in a sixth grade class. The average score for the whole class of a math test is 80. The average scores are 83 for boys and 78 for girls. How many girls are in the class?

21. Mr. Smith's age is half of his father's age. Mr. Smith's father's age is 15 times Mr. Smith's son's age. Two years from now, the sum of three people's ages will be 100. How old is Mr. Smith?

22. Three boxes are weighted by two at a time and the following weighs are obtained: 56 *kg*, 59 *kg*, and 60 *kg*. What is the difference in *kg* between heaviest box and the lightest box?

23. The sum of the number of students in class A and class B is 85. The sum of $\frac{3}{8}$ of students in class A and $\frac{3}{5}$ of students in class B is 42. How many students are in class A?

24. A store bought some oranges for re-sale. 40% were sold in the first day. 50% of the remaining was sold in the second day. The amount of oranges sold in the third day was $\frac{2}{3}$ of the amount sold in the first day. There were 50 *kg* of oranges left at the end of third day. How many kilograms of oranges were bought by the store for resale?

25. There are 60 students in a sixth grade class. Thirty nine of the 60 students are with the math club and 31 are with the computer club. Eight students are not in any of the two clubs. How many students are in both clubs?

26. You are given seven consecutive positive integers. Fill each small circle with one of the seven consecutive positive integers. The sum of any two numbers that are inside the two circles next to each other is the same and this sum is also the same as the number besides the line segment that connects the two circles. What integer should be in circle A?

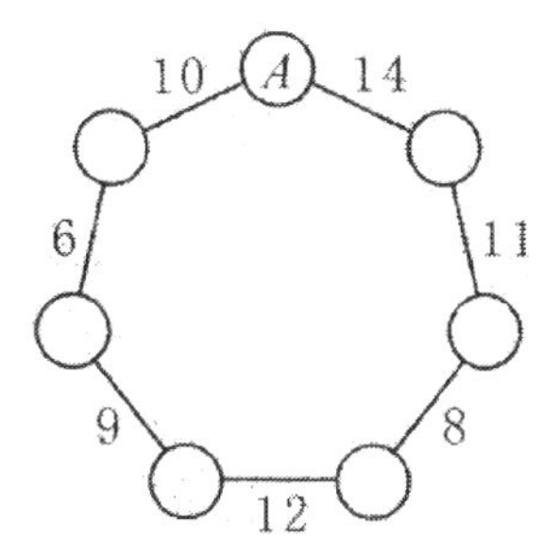

27. Three identical cups contain sweet water with the ratios 1:5, 1:8, and 1:9 of sugar to water, respectively. Now pour three cups into a larger container. What is the ratio of sugar to water of the mixture? Express your answer as a common fraction.

28. There are five cards each labeled with a different number of 1, 2, 3, 4, and 5. How many distinct sums can be obtained using these cards?

29. Sue owns 6 pairs of shoes and each pair has a distinct color. If she picks 4 shoes at random, what is the probability that she gets exactly one pair of the same color?

30. How many integers in $\{1, 2, 3, \ldots, 2011\}$ can be expressed as the difference of two square numbers?

Answer keys to practice 7

1. 1
2.123456
3. 61
4. 20 (A = 2010×2012)
5. 44
6. 316 (310+6)
7.54
8.9/28
9. 15
10. 21, 35
11. 46 cm
12. 5π
13.21 cm^2
14. 64.8 cm^2
15. 12π
16. 47 (38+9)
17. 10%
18. 300
19. 1000 g
20. 24
21. 30
22. 4 kg
23. 40
24. 1500 kg
25. 18
26. 6
27. 17/118
28. 13 (1 to 15, only 1 and 2 are not achievable).
29. 16/33 (C(6,1)*[C(10,2)-C(5,1)]/C(12,4))
30. 1508 (all the odd integers (1006) and multiples of 4 (502)).

.

MATHCOUNTS

■ Sprint Round Competition ■
Practice Test 8
Problems 1-30

Name

Date

DO NOT BEGIN UNTIL YOU ARE INSTRUCTED TO DO SO.

This round of the competition consists of 30 problems. You will have 40 minutes to complete the problems. You are not allowed to use calculators, books, or any other aids during this round. If you are wearing a calculator wrist watch, please give it to your proctor now. Calculations may be done on scratch paper. All answers must be complete, legible, and simplified to lowest terms. Record only final answers in the blanks in the right-hand column of the competition booklet. If you complete the problems before time is called, use the remaining time to check your answers.

Total Correct	Scorer's Initials

1. Calculate: $197 \times 198 - 196 \times 199$

2. How many two-digit prime numbers are there such that when the units and tens digits are switched, a new 2-digit prime number will be formed? 13 is one of the 2-digit prime numbers with this property.

3. There are 6 boxes of marbles and each box has the same numbers of marbles. If 200 marbles are taken from each box, the number of the remaining marbles is exactly the same as the sum of the number of marbles originally in two boxes. How many marbles are there in each box originally?

4. Four times the sum of n and 9 equal 8080. Find the value of n.

5. Mr. Smith and his 6th grade students are standing in a row to be counted. If the counting starts from the left side, Mr. Smith's number is 15. If the counting starts from the right side, Mr. Smith's number is 17. How many students are in Mr. Smith's class?

6. Alex was doing a math problem: first step a positive integer n was multiplied by 8, then 13 is subtracted. Alex made some mistake and what he did was the following: n was multiplied by 6 and then 3 was added. Fortunately Alex got the correct answer. Find the number n.

7. When a number n is taken out from the sum of some consecutive positive integers starting from 1, the new sum the numbers remaining is 265. What is n?

8. What is the greatest possible 3-digit positive integer that always has a remainder of 3 when it is divided by 4, 6, and 8?

9. List all the positive integers starting from1 that are not divisible by 3, what is the 100th number?

10. Add one number n to the list 1, 2, 3, 4, 5, 6, 7, 8, 9, and 10 such that the average value of the 11 numbers is 6. Find n.

11. There are 10 positive integers in a list. If the greatest number m is removed from the list, the average value of the 9 remaining numbers is 22. If the smallest number n is removed from the list, the average value of the 9 remaining numbers is 25. Find the value $m - n$.

12. Alex's class has 30 students. Fifteen of the 30 students are in the Math Club and twelve are in the Science Club. Six students in the class are in both clubs. m students are in only one club and n students are not in any of the two clubs. Find the values for $m + n$.

13. Several students put money together to buy some books. If each student pays \$9, there will be \$5 more than needed; if each student pays \$7, there will be \$9 less than needed. What is the cost of the books?

14. Alex and Sam run a 100-meter race. When Alex is 32 meters away from the finish line, Sam is 15 meters away from the finish line. When Alex is 20 meters away from the finish line, how many meters away from the finish line is Sam?

15. Alex and Bob together invest $500 for a business. If Bob adds $20, the amount of money invested by Alex will be exactly 3 times Bob's money. What is the amount of money Alex invests?

16. One day a ship goes with the current for 21 km and against current for 4 km. Another day the ship goes against the current for 7 km and then with the current for 12 km. The times used in the two trips are the same. Find the ratio of the speed of the ship to the speed of the current. Express your answer as a common fraction.

17. Five desks and 8 chairs cost $375. Each desk costs $10 more than each chair. Find the cost of each desk.

18. Alex runs exactly one round along a circular path of 400 meters. His speed is 6 meters per second for the first half of the time and 4 meters per second for the second half of the time. How many seconds does it take for Alex to run the second half of the distance of the path? Express your answer to the nearest seconds.

19. What is the acute angle formed by the hour and minute hands at 4:16 P.M.?

20. As shown in the figure, a square of side length 5 cm has some area overlapped with another square of side length of 4 cm. Find the difference of the non-overlapping areas of two squares.

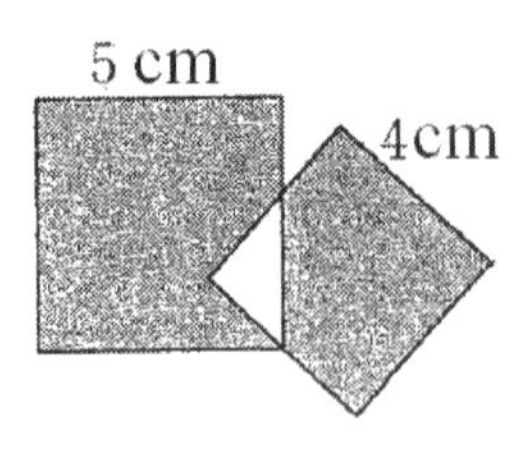

21. As shown in the figure, O is a point inside triangle ABC. OB bisects $\angle ABC$. OC bisects $\angle ACB$. $\angle BOC = 130°$. Find $\angle A$.

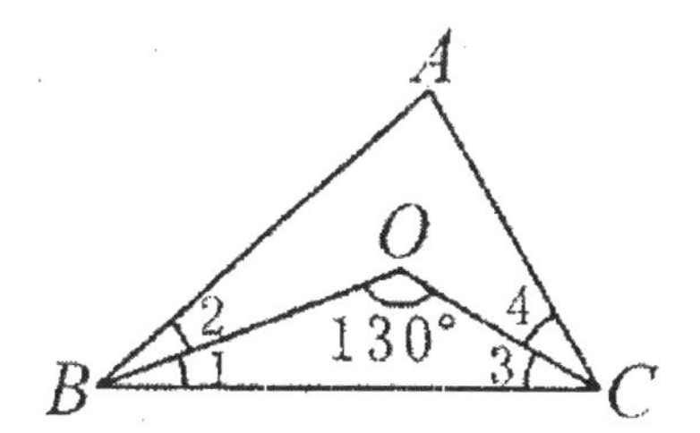

22. As shown in the figure, ABCD is a rectangle. 5ED = AD. 5BF = BC. 3AG = AB. 3HC = DC. EGFH is a parallelogram with the area of 21. Find the area of the rectangle ABCD.

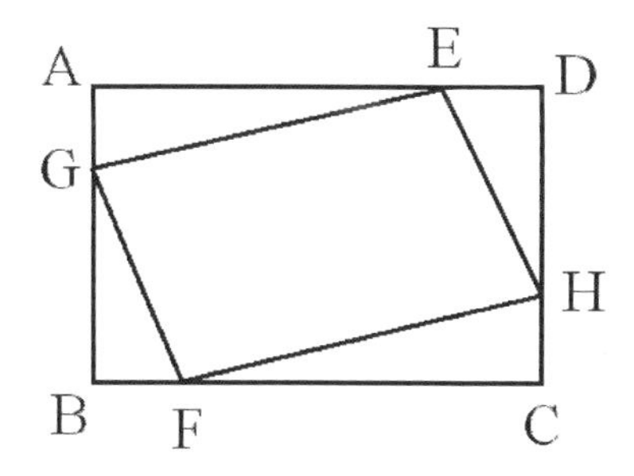

23. How many different 2-digit even numbers can be formed by using two digits from 0, 1, 2, and 3?

24. A 6-digit numbers in the form of $\overline{a2011b}$ is divisible by 45. What is the greatest sum of a and b?

25. Put 1, 2, 3, 4, 5, 6, 7, 8, and 9 in a row randomly. Any three neighboring digits can form a 3-digit number. There are 7 such 3-digit numbers for any one arrangement of these 9 digits. Find the smallest possible sum of seven such 3-digit numbers.

26. A dragonfly has 6 legs and 2 pairs of wings. A cicada has 6 legs and 1-pair of wings. A spider has 8 legs with no wings. The number of these insects is 60. The number of legs is 400. The number of wings is 50 pairs. Find the number of dragonflies.

27. There are 4 balls labeled 1, 2, 3, and 4, respectively, in a box. Randomly take out two balls and the positive difference of the numbers on these two balls is x. What is the greatest possible value of x?

28. Alex spent 6 days to finish reading a book of 300 pages. The number of pages he read in the third day was the sum of the numbers of pages he read in the first and second days. From the third day, the number of pages he read each day was the sum of the number of pages he read in the last two days. How many pages did Alex read in the fifth day?

29. Regular hexagon ABCDEF has an area of 24 square units. M and N are the midpoints of AF and CD, respectively. If MP//AB, MO//EF, PN//BC, and ON//ED, find the area of rhombus MPNO.

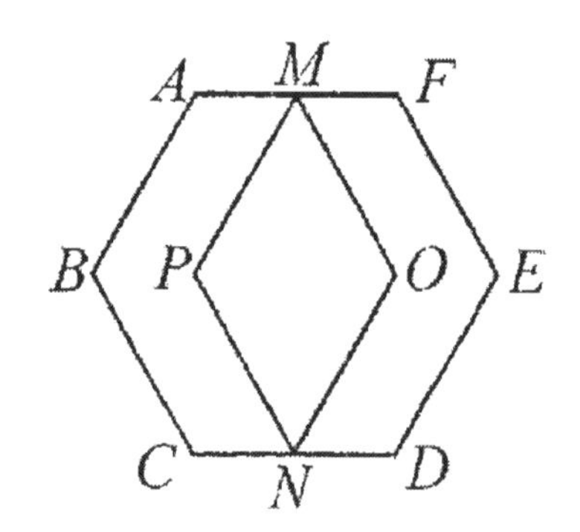

30. As shown in the figure, A, B, C, and D represent four 3-digit numbers, 863, 796, 375, and 483, but not necessarily in that order. Each symbol in the figure represents a different digit. What number is A?

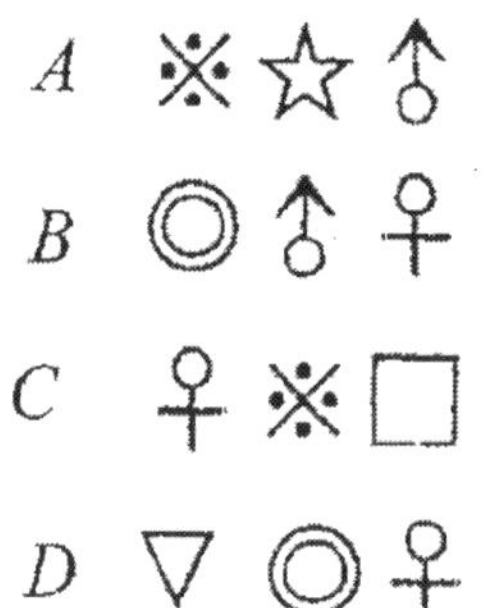

Answer keys to practice test 8:

1. 2
2. 8
3. 300
4.2011
5. 30
6. 8
7. 11
8. 987
9. 149
10. 11
11. 27
12. 24 (15+9)
13. $58
14. 0
15. 390
16. 2:1=2/1
17. 35
18. 46 2/3=47
19. 32°
20. 9
21.80°
22.35
23. 5
24. 9 + 5 = 14
25. 3122
26. 10
27. 1
28. 75
29. 8
30. 796

MATHCOUNTS

■ Sprint Round Competition ■
Practice Test 9
Problems 1-30

Name

Date

DO NOT BEGIN UNTIL YOU ARE INSTRUCTED TO DO SO.

This round of the competition consists of 30 problems. You will have 40 minutes to complete the problems. You are not allowed to use calculators, books, or any other aids during this round. If you are wearing a calculator wrist watch, please give it to your proctor now. Calculations may be done on scratch paper. All answers must be complete, legible, and simplified to lowest terms. Record only final answers in the blanks in the right-hand column of the competition booklet. If you complete the problems before time is called, use the remaining time to check your answers.

Total Correct	Scorer's Initials

1. Calculate $[223\times1.25+22.3\times75+2.23\times125]\times0.9+4$

2. Calculate:

$$1\frac{1}{2}+3\frac{1}{4}+5\frac{1}{8}+7\frac{1}{16}+9\frac{1}{32}+11\frac{1}{64}+13\frac{1}{128}+15\frac{1}{256}+17\frac{1}{512}+19\frac{1}{1024}=$$

Express your answer as a mixed number.

3. Find the greatest integer closest to S. $S=\frac{7}{20}+\frac{7}{20}\times2+\frac{7}{20}\times3+\cdots+\frac{7}{20}\times10$

4. Look at the pattern in the first three figures:

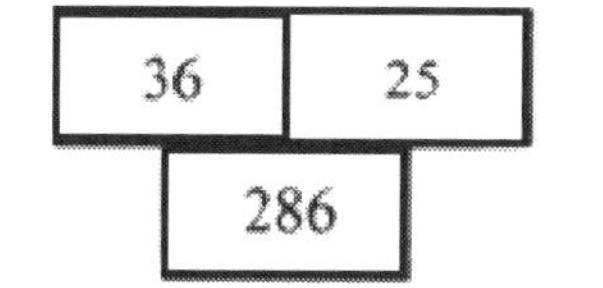

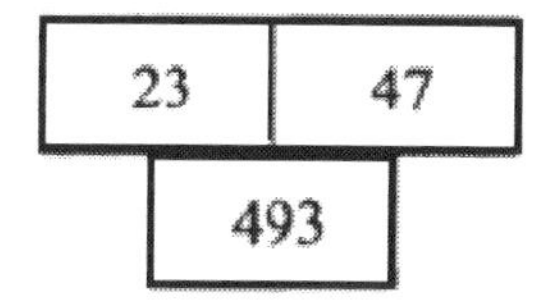

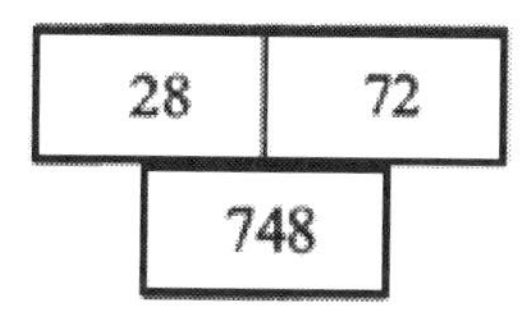

What number should go the blank rectangle in the following figure?

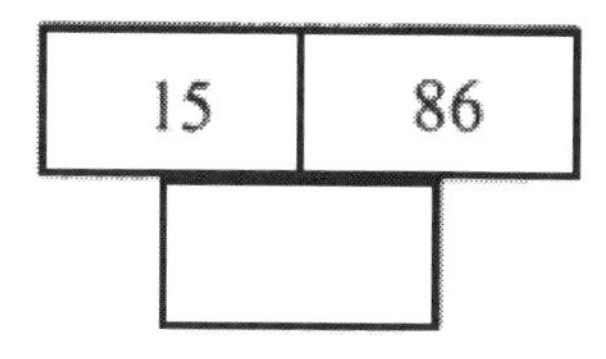

5. There are 10 red, 10 yellow, and 10 blue gloves in a bag. Each time you take out one glove from the bag without looking. At least, how many gloves do you need to take out to guarantee that you have at least two pairs of gloves with the different colors? If two gloves have the same color, they form a pair.

6. The sum of the digits of a 3-digit positive integer is 16. The units digit of the integer is bigger than the hundreds digit. If the positions of the units digit and hundreds digit are switched, a new number will be formed. The new number is 56 more than two times the original number. Find the original 3-digit number.

7. There are 150 2×2 tiles. How many different rectangular shapes can be obtained using these 150 tiles?

8. If two consecutive odd numbers are both prime numbers, they are called twin prime numbers. (3, 5) and (5, 7) are examples of twin prime numbers. How many pairs of twin prime numbers are there less than 100?

9. The product of Alex, Bob, and Charles' ages is n. The product of n and 361 is 361361. What is Alex's age if he is the oldest among the three people?

10. Fill each square with a different digit to form two 2-digit positive integers and make the following equation true. What is the greatest sum of the four digits used?

$$\square\square \times \square\square = 1995$$

11. The product of two different positive integers is 7 times their sum. Find the larger one of the two numbers.

12. All six graders in Hope Middle School went to a math competition. The number of boys is 4/9 of the number of total students. The sum of the scores obtained by boys is 6/7 of the sum of the scores obtained by girls. If the average score for boys is 90, find the average score for girls.

13. Alex and Bob start at the same time to climb a hill. When they reach the top of the hill, they turn back immediately and double their speeds. By the time Alex reaches the top of the hill, Bob is 500 meters away from the top of the hill. When Alex returns to the bottom of the hill, Bob is still at the middle of the hill. Find the distance from the bottom to the top of the hill.

14. The hour hand and the minute hand are in the opposite position after 4 o'clock. What time is it?

15. $\overline{k45k7}$ is a 5-digit positive integer divisible by 3. The number of possible values for k is m. The number of such 5-digit positive integers that are also divisible by 9 is n. Find the value of $m + n$.

16. Seven lattice points are marked in the figure. Use these 7 points as the vertices to from some squares and isosceles right triangles. The number of squares formed is m and the number of isosceles right triangles formed is n. Find the value of $m + n$.

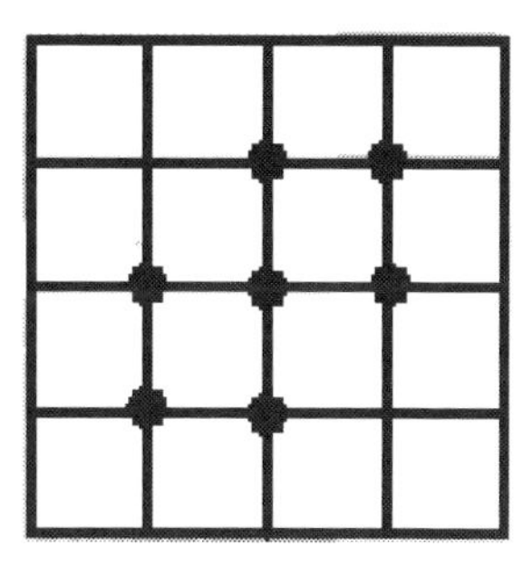

17. Five times the number of girls in a class is the same as four times the number of boys in the same class. If 10 girls leave and 30 boys leave, two times the number of girls remaining is 7 more than three times the number of boys remaining. How many girls are in the class originally?

18. Find the 89th term in the sequence: 2, 9, 17, 24, 32, 39, 47, 54, 62, …

19. Four high school boys and four girls are assigned to four middle schools to proctor the AMC 8 exam. Each middle school needs to have one boy and one girl. How many ways are there to assign them?

20. A frog is jumping from point A in the clockwise direction as shown in the figure. First, the frog jumps 1 step, say jumping from A and landing on B. Second, the frog jumps 2 steps, say jumping from B passing C and landed on D. Third, the frog jumps 3 steps, say jumping from D and landed on A. Continuing the pattern until it jumps a total of 2011 steps. How many of these six vertices of the hexagon have the frog not landed on?

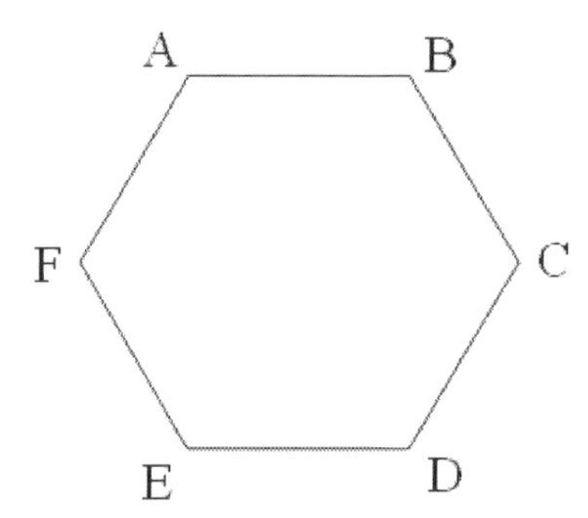

21. ABCD is a square. AD = 4 cm. BE = 2 cm. and DF = 1 cm. Find the shaded area. Express your answer as a mixed number.

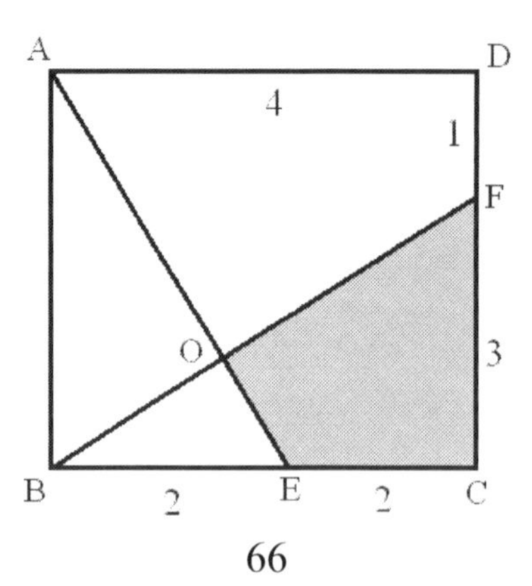

22. Car A leaving Kinston travels to Durham and Car B leaving Durham travels to Kinston at the same time along the same road. They meet at the place that is 60 km from Durham. Both cars turn around right way after they arrive at their destinations, and they meet again at the place that is 50 km from Kinston. Find the distance from Kinston to Durham.

23. How many palindrome numbers are there from 1 to 2011?

24. Find the sum of the last four digits of S if

$S = 15 + 195 + 1995 + 19995 + \ldots + 1\underbrace{9999..99}_{\text{repeated 44 times}}5$

25. What is the 100th digit in the list: 182764125216...... ?

26. *ABC* is a triangle. *D* is a point on *AB*. Connect *CD*. *E* and *F* are two points on *CD*. The areas of ΔADE, ΔAEC, ΔBDF, and ΔBFC are $S_1 = 1$, $S_2 = 4$, $S_3 = 7$, and $S_4 = 3$, respectively. Find the area of ΔBEF.

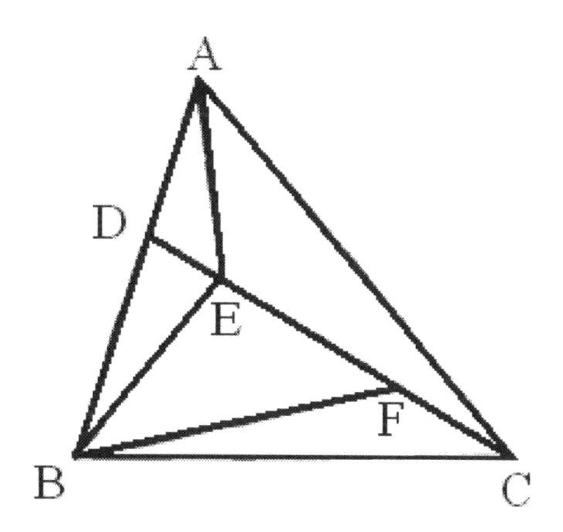

27. What is the last digit of the greatest value of k such that 3^{2011} can be expressed as the sum of k consecutive positive integers?

28. Two cars drive in the same direction. One is driving at 40 mph and the other is driving at 50 mph. They are 15 miles apart. In how many hours will the cars be exactly 30 miles apart? Express your answer as a decimal to the nearest tenth.

29. A rectangular box has the length 3 cm, width 2 cm, and height 1 cm. When the rectangular box is unfolded, what is the smallest possible perimeter of the plane figure (net)?

30. ABC is an isosceles right triangle. AB = AC = 2 cm. Use B and C as the centers to draw arcs AD and AE. Find the shaded area. Express your answer in terms of π.

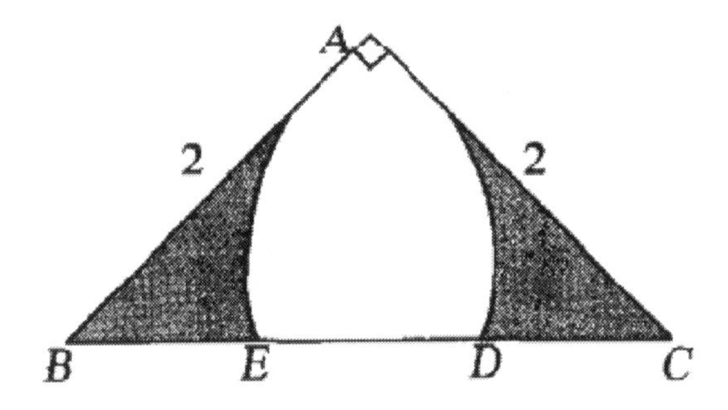

Answer keys to practice test 9

1. 2007+ 4=2011
2. $100\frac{1023}{1024}$
3. 19
4. 875
5. 13
6. 439.
7. 6
8. 8 pairs
9. 13
10. 20
11. 56
12. 84
13. 3000 m
14. $4:54\frac{6}{11}$
15. m + n = 3+1 =4
16. 3+14=17
17. 36
18. 662
19. 576
20. 2 (C and F)
21. $4\frac{10}{11}$ cm^2.
22. 130 km.
23. 119.
24. 24
25. 2
26. 5
27. 6.
28. 1.5 hour
29. 22 cm
30. 4-π

MATHCOUNTS

■ Sprint Round Competition ■
Practice Test 10
Problems 1-30

Name

Date

DO NOT BEGIN UNTIL YOU ARE INSTRUCTED TO DO SO.

This round of the competition consists of 30 problems. You will have 40 minutes to complete the problems. You are not allowed to use calculators, books, or any other aids during this round. If you are wearing a calculator wrist watch, please give it to your proctor now. Calculations may be done on scratch paper. All answers must be complete, legible, and simplified to lowest terms. Record only final answers in the blanks in the right-hand column of the competition booklet. If you complete the problems before time is called, use the remaining time to check your answers.

Total Correct	Scorer's Initials

1. Calculate: $2011 \div 37 + 270 \div (37 \times 2)$

2. Calculate: $82.54 + 835.27 - 20.38 \div 2 + 2 \times 6.23 - 390.81 - 9 \times 1.03$

3. In a 6th grade class, the number of girls is two times the number of boys. If the average height is 150 cm for girls and 162 cm for boys, what is the average height of all students?

4. How many triangles are there?

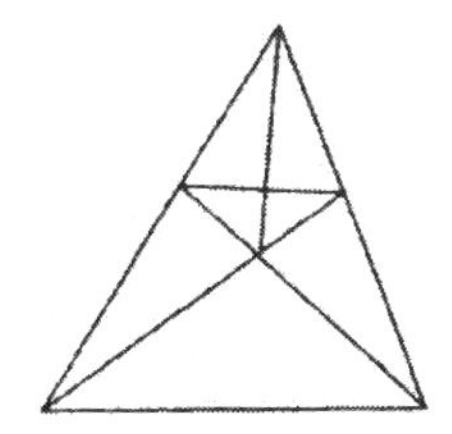

5. Mr. Edwards has the same number of oranges, apples and bananas. He gives each student 1 orange, 3 apples, and 5 bananas. He has 24 apples left. What is the sum of the numbers of apples and bananas left?

6. Alex arranged some red balls in a row on a table. He then put two yellow balls between every two red balls. Finally he put two blue balls between any two balls that are already on the table. Once he found a pattern, he continued the pattern until the total number of balls on the table was 2011. How many balls were yellow?

7. How many 3-digit positive integers are there such that each digit is different from other digits and the sum of the digits is 6?

8. There are three boxes containing apples. If they are weighted by two at a time, the weights are 63 kg, 65 kg, and 66 kg, respectively. What is the difference between the heaviest box and the lightest box?

9. Alex has an infinite amount of coins valued at 1¢, 2¢, 5¢, and 10¢. He takes out 16¢ to buy 1 pencil with no change needed. At most how many different ways are there?

10. Fill each square with one of the digits 1, 2, 3, 4, and 5. What is the greatest possible product?

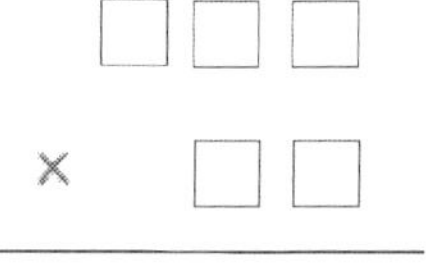

11. What is the units digit of the 2011th number in the following Fibonacci sequence 1, 1, 2, 3, 5, 8, 13, 21, 34,…,?

12. Fill each square with one of the digits 1, 2, 3, 4, and 5 such that the digit in the black square is bigger than the two neighboring digits. How many different arrangements are there?

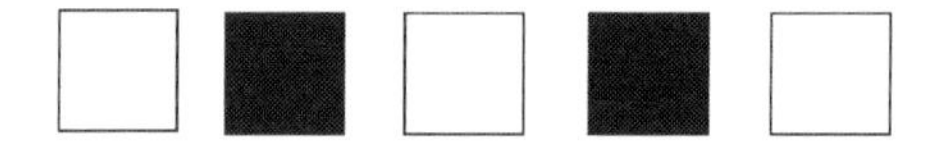

13. As shown in the figure, the line segments represent the roads of a park consisting of 8 congruent right triangles. The long leg of each triangle is 240 meters and the short leg is 100 meters. The hypotenuse is 260 meters. Alex likes to walk through every road of the park, starting at point A and ending also at A. If he keeps his walking speed at 60 meters per minute, what are the least possible number of minutes he needs?

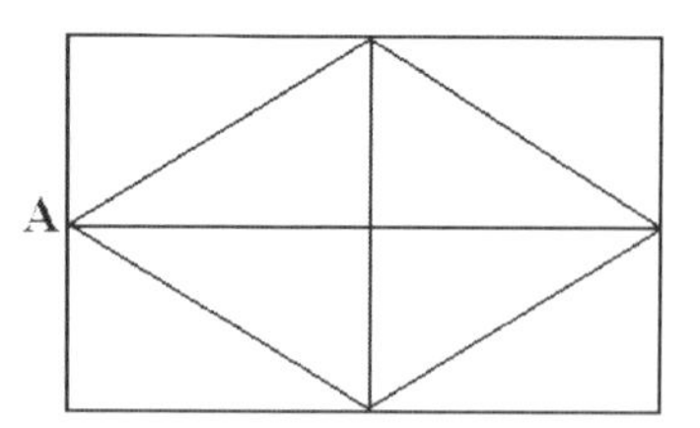

14. Yesterday 420 apples were evenly divided among a 6th grade class. Today a new student comes and 420 apples are still evenly divided but each student gets two apples less than last time. How many students were there yesterday?

15. As shown in the figure, a rectangle with the area of 2009 cm^2 is divided into one smaller rectangle, two isosceles right triangles, and three trapezoids. The isosceles right triangle and the trapezoid have the same area. B is the midpoint of AC. Find the shaded area.

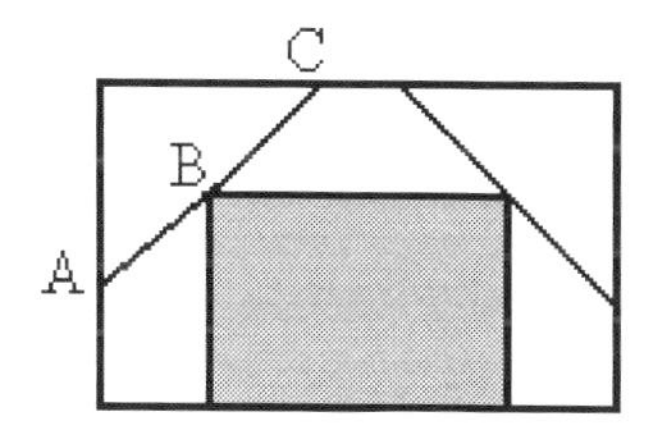

16. Using the digits 4, 5, 6, 7, 8, and 9 each once to form a 6-digit number that is divisible by 667. What is the quotient when the 6-digit number is divided by 667?

17. Select some numbers from positive integers 1 to 6 such that the sum of the selected numbers is a multiple of 3, but not a multiple of 5. How many different ways are there to do this? Note that you are allowed to select one number.

18. If books are put into boxes with 24 books in each box, the last box will be 2 books short. If the books are put into boxes with 28 books in each box, the last box will still be 2 books short. If the books are put into boxes with 32 books in each box, the last box will contain 30 books. How many books are there if the number of books is less than 1000?

19. If congruent regular pentagons are used to form a circular shape, how many pentagons are needed? The following figure shows part of the circular shape with 3 pentagons.

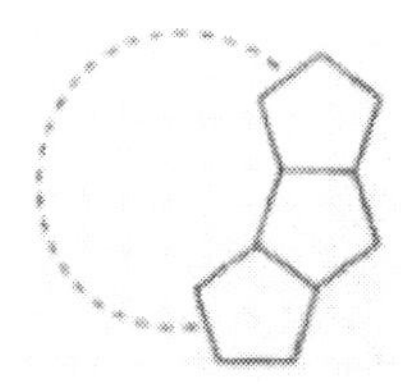

20. The product of the digits of a 5-digit positive integer is 25 times the sum of its digits. What is the greatest possible value of the 5-digit number?

21. The area of a regular hexagon $A_1A_2A_3A_4A_5A_6$ is 2009 cm^2. B_1, B_2, B_3, B_4, B_5, and B_6 are the middle points of the sides of the hexagon as shown in the figure. Find the shaded area.

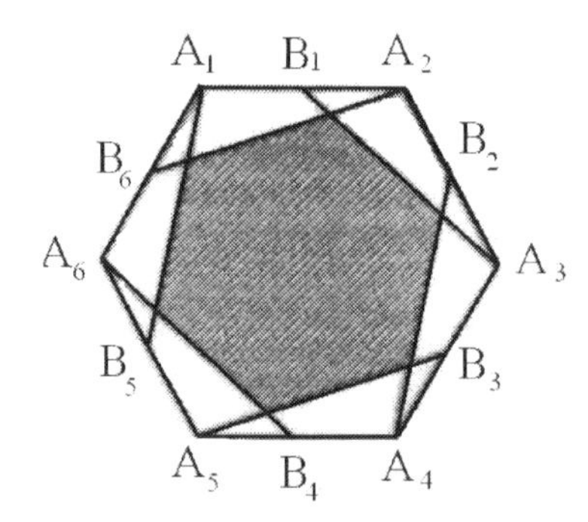

22. There are 64 $1\times1\times1$ cubes. Thirty four of them are white and thirty are black. A $4\times4\times4$ cube is formed using these 64 small cubes. What is the greatest ratio of the white area to the black area of the surface of the $4\times4\times4$ cube? Express your answer as a common fraction.

23. A mail man needs to go through every street to deliver mail starting from the post office and coming back after he finishes. What is the shortest distance he needs to go? The line segments represent the street and the number on each line segment is the length of the street in miles. He is allowed to go through any street more than once. All streets are two-way streets.

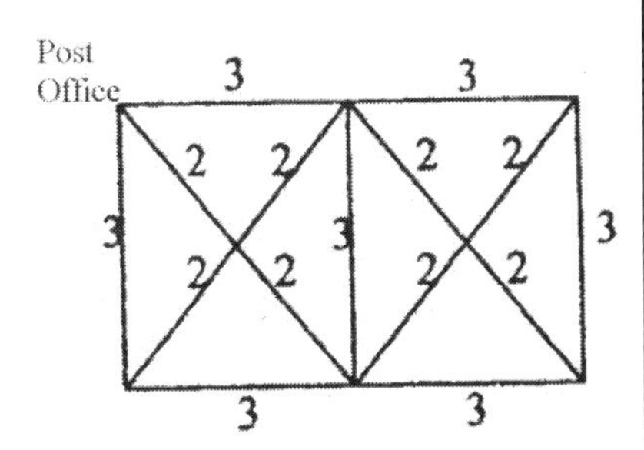

24. There are 12 teams in a soccer tournament. Every two teams play exactly once. The winner gets 3 points and the loser gets 0 points. One point is given to both teams for a tie game. When the tournament is over, what is the greatest possible difference between the 3rd team's score and the fourth team's score?

25. Some of the faces of one N×N×N cube are painted red and then cut into 1×1×1 cubes. The number of cubes with at least one face painted is 52% of the number of total 1×1×1 cubes. Find the value of N.

26. If two composite numbers are relatively prime and their least common multiple is 126, find their sum.

27. $\overline{AB}$ is a 2-digit positive integer with two distinct digits and has the following property: $(\overline{AB})^2 - (\overline{BA})^2 = x^2$. Find $\overline{AB}$.

28. Calculate: $1155\times(\frac{5}{2\times3\times4}+\frac{7}{3\times4\times5}+\cdots+\frac{17}{8\times9\times10}+\frac{19}{9\times10\times11})$

29. A special clock is special in the following way:

(1). The minute hand completes one round every 10 minutes; and (2). While the minute hand moves 6 rounds, the hour hand only moves one round.

Now both hands are overlapped. We call this the first time they lie in a straight line. How many minutes does it take for them to lie in a straight line 6 times?

30. ABCD is a parallelogram. Triangle ABP has an area of 73 cm^2 and triangle BPC has an area of 100 cm^2. Find the area of triangle BPD.

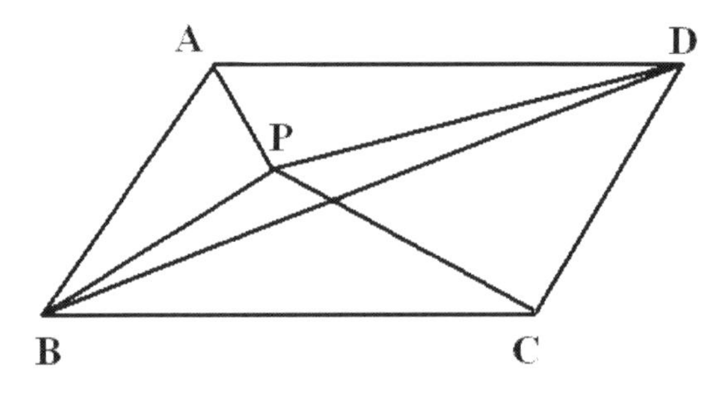

Answer keys to practice test 10:

1. 58.
2. 520.
3. 154 cm.
4. 20.
5. 48.
6. 447.
7. 14.
8. 3 kg.
9. 25.
10. $52 \times 431 = 22412$
11. 9. ($2011 = 60 \times 33 + 31$.The pattern repeats every 60 digits)
12. 16.
13.60 min.
14.14.
15. 861 cm^2.
16.1434.
17. 19.
18. 630 books.
19. 10.
20. 55332.
21. 1148 cm^2.
22. 37/11.
23. 46 miles.
24. 22.
25. 5.
26. 23.
27. 65.
28. 651
29. 30 min.
30. 27 cm^2.

MATHCOUNTS

■ Sprint Round Competition ■
Practice Test 11
Problems 1-30

Name

Date

DO NOT BEGIN UNTIL YOU ARE INSTRUCTED TO DO SO.

This round of the competition consists of 30 problems. You will have 40 minutes to complete the problems. You are not allowed to use calculators, books, or any other aids during this round. If you are wearing a calculator wrist watch, please give it to your proctor now. Calculations may be done on scratch paper. All answers must be complete, legible, and simplified to lowest terms. Record only final answers in the blanks in the right-hand column of the competition booklet. If you complete the problems before time is called, use the remaining time to check your answers.

Total Correct	Scorer's Initials

1. $\frac{1}{2}+\frac{14}{28}+\frac{104}{208}+\frac{1004}{2008}=$______

2. If $a*b=a+b\div a$, find $(1*2)*3=$_______

3. If $3A=4B=5C$, what is $A:B:C=$

4. The list 1, 3, 9, 25, 69, 189, 517, … has the following property: starting from the third number in the list, each number is 1 more than 2 times the sum the two numbers before it. For example, $9 = 1+2\times(1+3)$. What is the remainder when the 2011th number in the list is divided by 6?

5. Three days are used to fish and two days are used to dry the fishing net. Under this timing pattern, how many days would you fish in 100 days?

6. Alex wants to calculate the average value of six numbers. He should use division in the last step, but instead he uses multiplication. His wrong result is 1800. Find the correct answer for Alex.

7. A 3-digit number $\overline{abc}$ is 99 less than another 3-digit number $\overline{cba}$. If a, b, and c are all distinct, what is the greatest value for $\overline{abc}$?

8. The total number of apples in two bags is 20. The number of apples in each bag is the same if 7 apples are taken out of bag A and put into bag B. How many apples are in bag A originally?

9. The following figure is a monthly calendar in March. The sum of the digits of the four numbers inside the rectangle is 5+6+1+2+1+3=18. What is the greatest sum of the digits of four numbers that can be obtained by moving this rectangle around the calendar?

						1
2	3	4	5	6	7	8
9	10	11	12	13	14	15
16	17	18	19	20	21	22
23	24	25	26	27	28	29
30	31					

10. During Thanksgiving, a store uses the following discounts for an electrical appliance originally marked $100:

(1) Take $20 off the original price and then deduct 20%.

(2) Discount by 20% of the original price, and then take $20 off.

What is the positive difference of money a customer can have from these two ways?

11. ABCD is a square with AB = 12 cm. Point E is on CD. $BO \perp AE$ at O. $OB = 9$ cm. How long is AE?

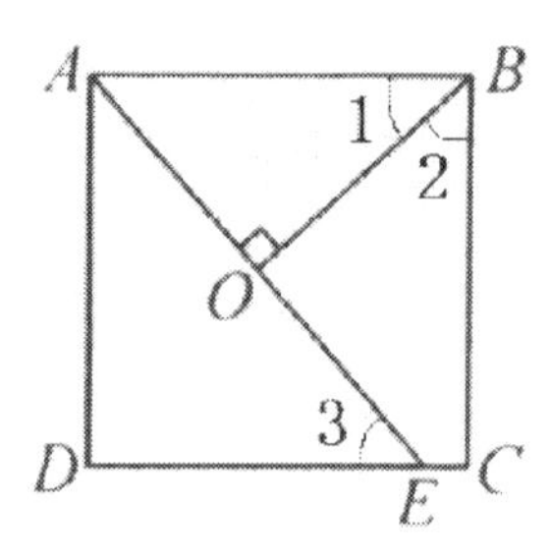

12. As shown in the figure, the length of one side of each small square is 1 cm. How many different triangles with an area of 3 cm^2 can be drawn using a combination of three of the lattice points as vertices?

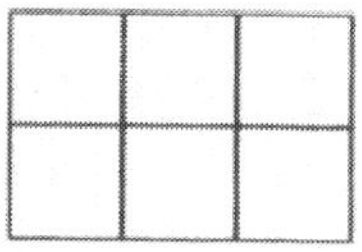

13. There are 27 seats in a row and n seats already taken. Alex finds that no matter which seat he is going to take, he is always beside someone. Find the least value of n.

14. How many ways are there to take three numbers from 1 to 9 such that the sum of these three numbers is a multiple of 3?

15. There are 4 red balls, 7 yellow balls, and 8 black balls in a bag. At least how many balls are needed to be taken out randomly in order to guarantee getting 6 balls of the same color?

16. There are 6 bags of candy on the table containing 3, 4, 5, 7, 9, and 13 pieces of candy, not necessarily in that order. Alex takes away 2 bags, and Bob takes away 3 bags. The number of candy Bob has is 2 times of the number of candy Alex has. What is the number of candy in the bag left on the table?

17. Two years ago, Alex's age was 4 times the age of his son, Bob. Two years from now, Alex's age will be 3 times Bob's age. How old is Alex now?

18. Alex has some toy cars and toy airplanes. The total number of cars and airplanes is 30. Each toy airplane has three wheels and each toy car has 4 wheels. The total number of wheels is 110. How many toy airplanes does Alex have?

19. If $3a+2b=24$, find the value of $\frac{3}{4}a-5+\frac{1}{2}b$.

20. $6.90 can buy 3 pounds of apples and 2 pounds of oranges. $22.80 can buy 8 pounds o apples and 9 pounds of oranges. How much does it cost to buy 1 pound of apples and 1 pound of oranges?

21. ABCD is a square. AB = 4. BEFG is also a square. BE =6. O_1 is the center of ABCD and O_2 is the center of BEFG. Find the shaded area O_1 O_2 B.

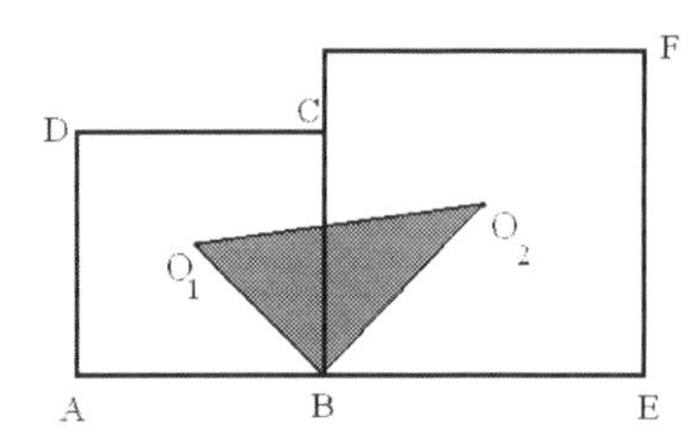

22. What is the angle formed by the minute and hour hands at 4:16 P.M.?

23. If $\frac{1}{2072}+\frac{1}{65009}=\frac{1}{A}$, find the value for A.

24. Alex can finish a job alone in 40 days. If Bob works on the job for 10 days and Alex joins in on the eleventh day, they can complete the job by working together for another 20 days. How many days will it take for Bob to finish the job if he works alone?

25. You are given a 3-digit positive integer. When you reverse the digits of the 3-digit positive integer, you still get a 3-digit positive integer. The sum of these two 3-digit positive integers is 888. How many such 3-digit positive integers are there?

26. There are 7 cards each with a number on it: [1]、[1]、[2]、[3]、[9]、[9]、[9]. How many ways can you take 3 cards to make a 3-digit even integer? Note that card [9] can be treated as [6] if flipped around.

27. The circle graph shows the results of a 2010 family spending of $24,300. How many dollars were spent on Education?

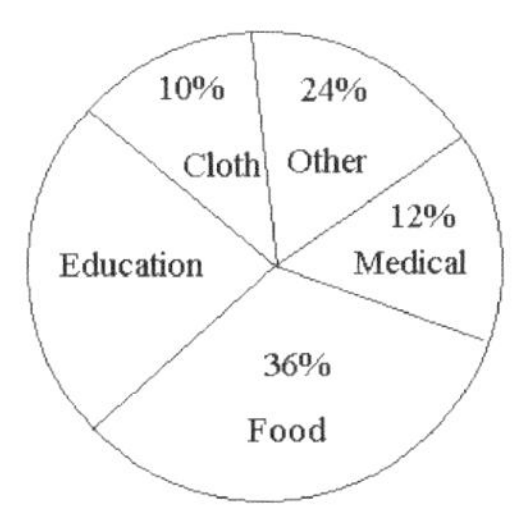

28. Point O is on the line segment AB. $\angle BOC$ is 90°.
$\angle BOD : \angle COD = 4:1$. Find $\angle AOD$.

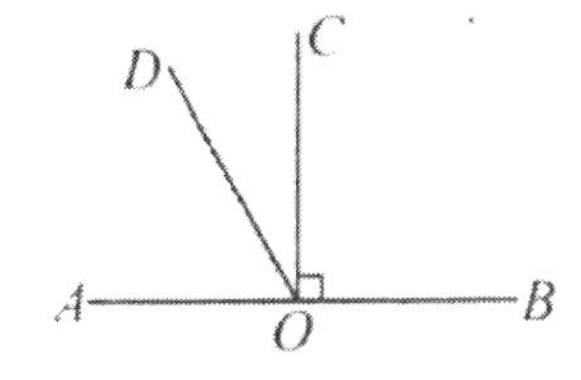

29. Alex has 6 red balls and 4 green balls. How many ways are there such that Alex can arrange them in a row so that no two green balls are next to each other?

30. A group of monkeys can evenly divide 56 pears among themselves (every monkey gets the same number of pears). If four more monkeys join them and they have to re-do

the distribution, each monkey will get n pears and there will be one pear left over. Find the value of n.

Answer keys to practice test 11:

1. 2.

2. 1*2=1+2÷1=3, 3*3=3+3÷3=4

3. $A:B:C=\frac{1}{3}:\frac{1}{4}:\frac{1}{5}=20:15:12$

4. Note the pattern of the remainders when these numbers is divided by 6:

1, 3, 3, 1, 3, 3, 1, 3, 3, ... Since 2011÷3=670...1, the remainder is 1.

5. 3 days fishing every 5 days. 100÷5×3=60

6. 1800÷6=300. 300÷6=50

7. $\overline{abc}+99=\overline{cba}$, $a=c+9$, To make $\overline{abc}$ greatest, if $a=9$, then $c=0$, $\overline{cba}$ is not 3-digit number, contradiction. If $a=8$, then $c=9$, b=7, so $\overline{abc}$ =879。

8. (20+7×2) ÷2=17

9. Group is 18、19、25、26. 1+8+1+9+2+5+2+6=34

10. 0.8a-16- 0.8a-20= $4.

11. In OECB, ∠2+∠OEC=180° , since∠3+∠OEC=180° , so∠2=∠3, ∠1=∠DAC, so, $\frac{AB}{AE}=\frac{OB}{AD}$, that is $\frac{12}{AE}=\frac{9}{12}$. Answer: $AE=16$

12. 6=1×6=2×3. when the base is 3, we have 4×2=8. When the base is =2, we have 1×2=2. So total 8+2=10.

13. 9

14. (1) All 3 numbers are multiple of 3: 1 way

(2) the remainder is 1 when divided by : 1way

(3) the remainder is 2 when divided by 3: 1 way

(4) one number has remainder 1, one number has the remainder 2 and one number has the remainder 3: 3×3×3=27ways. Total 1+1+1+27=30 ways.

15. 4+5+5+1=15

16. 13+7+4=2× (3+9) , 5.

17. $2\times(x+4)=3x$, $x=8$, Alex's age 8×4=32. This year 32+2=34.

18. 10

19. $\frac{3}{4}a-5+\frac{1}{2}b=\frac{1}{4}(3a+2b)-5=1$.

20. 29.7/11=2.7.

21. The difference between the areas of one right trapezoid minus two right trapezoid. (2+3)×5÷2-2×2÷2-3×3÷2=6.

22. 120-6×16+0.5×16=32.

23. $\frac{1}{2072}+\frac{1}{65009}=\frac{1}{8\times7\times37}+\frac{1}{7\times37\times251}=\frac{1}{259}\times\frac{259}{2008}=\frac{1}{2008}$, so A=2008.

24. 60.

25. 7

26. 15 (units digit is 2) +23 (units digit is 6) =38

27. 24300× (1-10%-24%-12%-36%) =4374.

28. $\angle BOD:\angle COD=4:1$, so $\angle BOC:\angle COD=3:1$. $\angle COD=30°$. $\angle AOD$=60.

29.35.

30. 56 has the factor: : 1, 2, 4, 7, 8, 14, 28, and 56. 55 has the factors:1, 5, 11, and 55. Only 11=7+4, so there are originally 7 monkeys, Answer:55÷11=5.

MATHCOUNTS

■ Sprint Round Competition ■
Practice Test 12
Problems 1-30

Name

Date

DO NOT BEGIN UNTIL YOU ARE INSTRUCTED TO DO SO.

This round of the competition consists of 30 problems. You will have 40 minutes to complete the problems. You are not allowed to use calculators, books, or any other aids during this round. If you are wearing a calculator wrist watch, please give it to your proctor now. Calculations may be done on scratch paper. All answers must be complete, legible, and simplified to lowest terms. Record only final answers in the blanks in the right-hand column of the competition booklet. If you complete the problems before time is called, use the remaining time to check your answers.

Total Correct	Scorer's Initials

1. Calculate: $13 \times 23 + 1001$

2. A rectangle with the length 12 cm and width 10 cm is cut into two parts along the dashed line as shown in the figure. What is the sum of the perimeters of the two parts?

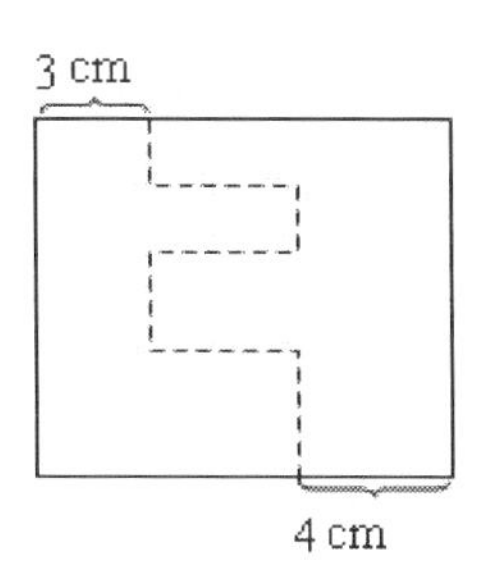

3. There are 10 rows of the following arrangement of 3's. What is the sum of all numbers?

3

3 3 3

3 3 3 3 3

……

4. The average age of five teachers is 36 and no one is less than 30 years old. Find the greatest possible age of the oldest one.

5. Find $143 \div ab$ if $7a = 11$ and $9b = 13$.

6. Alex and Bob are standing in front of a big mirror. Alex looks at the mirror and reads the number on Bob's sportswear as shown in the figure. What is the number of Bob's sportswear?

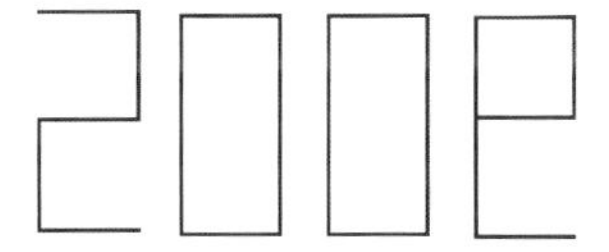

7. An unlimited number of darts are to be thrown at a dartboard with possible scores as shown in the figure. At least how many darts must be thrown in order to get exactly 120 points?

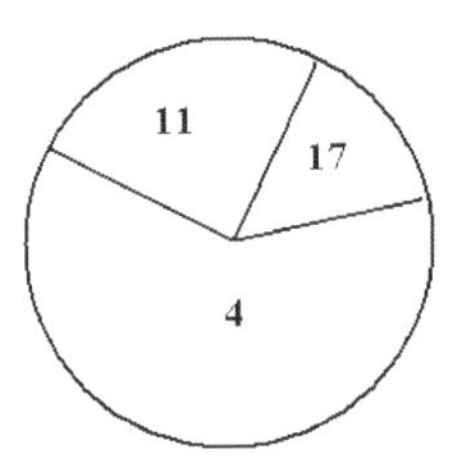

8. What is the smallest units digit of the product of three consecutive odd positive integers?

9. A 2-digit positive integer has 5 divisors. Find the greatest value of such a 2-digit number.

10. How many integers from 1 to 1000 are multiples of 2 and 3 but not 5?

11. A 6th grade class has 23 students. The average score of their math midterm test was 72. A new student named Sam with a score of 96 is added to the class. What will be the new average score for the class?

12. How many 3-digit numbers divisible by 6 are there such that when the digits are rearranged, the new numbers are still divisible by 6?

13. m candies are distributed between n kids. If each kid gets 3, there will be 24 candies left. If each kid gets 5, there will be 14 left. Find $m + n$.

14. Alex's age is 7 times his grandson Bob's age. Alex's age was 10 times Bob's age 3 years ago. Alex's age was n times Bob's age 6 years ago. Find n.

15. Peter and Sam together have 24 candies. If Peter eats 3 of his candies, and Sam eats 2 of his candies, then the number of Peter's candies will be 3 more than the number of Sam's candies. How many candies does Peter have originally?

16. A square-shaped pond has the side length of 100 meters. 208 trees are planted two meters away from the pond also in a square shape. If the trees are evenly spaced, find the distance between any two trees.

17. Five machines can produce 600 parts in 3 hours. How many parts can 11 machines produce in 8 hours?

18. Alex walked to school this morning. His brother Bob left on a bike 30 minutes later to catch him. Bob's speed is two times Alex's speed. How many minutes did it take for Bob to catch Alex?

19. In a country there are two kinds of coins valued at 2¢ and 5¢. You are a visitor of that country and you have ten coins of each kind. How many different payments can you make by using these coins?

20. Hope Middle School starts their first class at 8:00 A.M. There are 15-minute breaks between every two periods. There are 4 periods in the morning with each period lasting 40 minutes. What time is it when 4th period is finished?

21. ABCD is a square with the area of 36. DE = 4. Find the length BF.

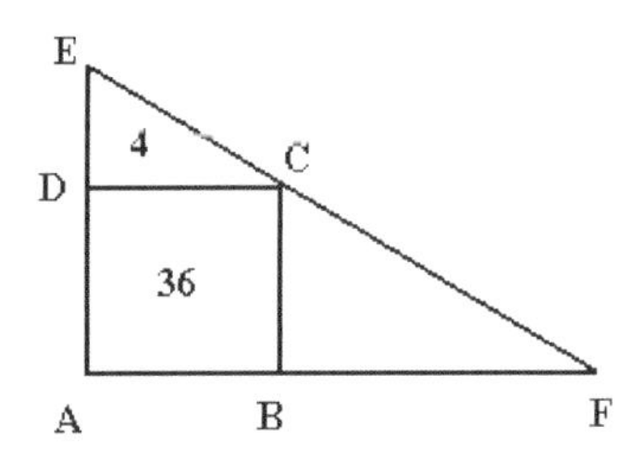

22. As shown in the figure, three sides of triangle ABC are extended. Find $\angle 1 + \angle 2 + \angle 3$.

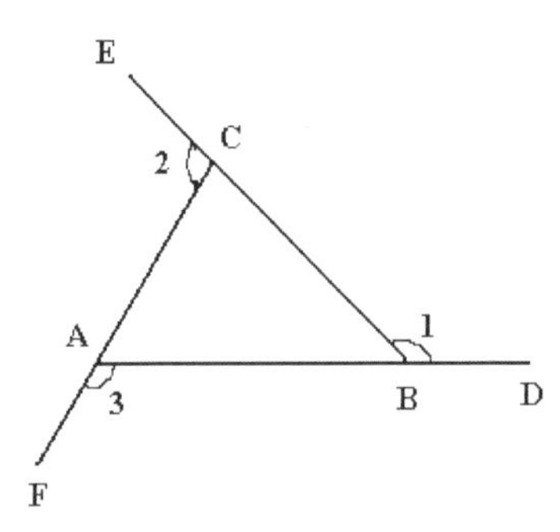

23. What is the greatest possible value of the remainder when 2011 is divided by a two-digit number?

24. There are m ways to take two fruits from 5 apples, 3 oranges, and 2 bananas. There are n ways to take two different fruits from 5 apples, 3 oranges, and 2 bananas. Find $m + n$.

25. 5-digit number $\overline{8abc8}$ is divisible by 2009. Find 3-digit number $\overline{abc}$.

26. The units digit of a 6-digit number is 1. If the digit 1 is moved to the leftmost position, the new number is one-third of the original number. Find the original 6-digit number.

27. A bag contains 9 red balls, 6 yellow balls, 2 green balls, and one purple ball. These balls are identical except different in color. At least how many balls must be taken out of bag without looking to guarantee that 4 balls are the same color?

28. Alex and his four cows want to cross a river. The times taking for cows A, B, C, and D to cross the river are 1 minute, 2 minutes, 5 minutes, and 6 minutes, respectively. Alex needs to go with two cows when crossing the river (he rides on the back of one of the two cows). He needs to ride on the back of one cow for returning. At least how long does it take for Alex and his cows to cross the river?

29. Alex has 7 different small gifts. He likes to choose five of the seven gifts for his five friends Bob, Catherine, Danny, Emily, and Frank with one gift for each person. Bob likes to have the puzzle book or the toy car. Catherine likes to have the camera or the puzzle book. How many ways are there for Alex to give out the five gifts?

30. As shown in the figure, both ABCD and BEFG are squares. The shaded area is 10. Find the area of ABCD.

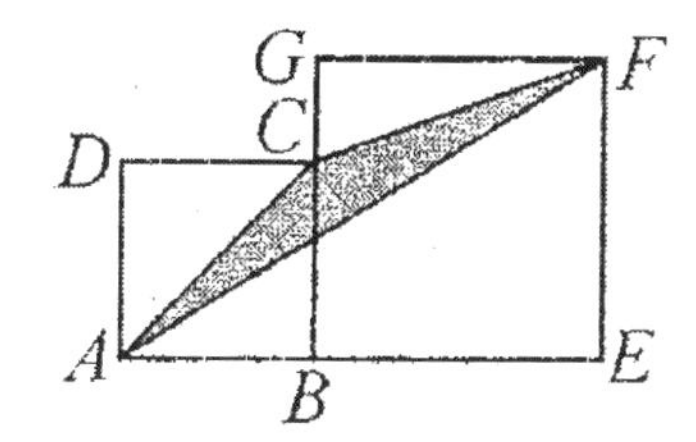

Answer keys to practice test 12

1. 1300
2. 94 cm
3. 300
4.60
5. 63
6. 9005
7. 10 (17×4+11×4+4×2)
8. 3
9. 81
10. 133
11. 73
12. 16
13. 44
14. 19
15. 14
16. 2
17. 3520
18.30 min.
19. 66
20. 11:25
21. 9
22.360°
23. 91
24. 31+45=76
25. 437
26. 428571
27. 10
28. 13 min.
29. 180
30. 20

MATHCOUNTS

■ Sprint Round Competition ■
Practice Test 13
Problems 1-30

Name

Date

DO NOT BEGIN UNTIL YOU ARE INSTRUCTED TO DO SO.

This round of the competition consists of 30 problems. You will have 40 minutes to complete the problems. You are not allowed to use calculators, books, or any other aids during this round. If you are wearing a calculator wrist watch, please give it to your proctor now. Calculations may be done on scratch paper. All answers must be complete, legible, and simplified to lowest terms. Record only final answers in the blanks in the right-hand column of the competition booklet. If you complete the problems before time is called, use the remaining time to check your answers.

Total Correct	Scorer's Initials

1. Calculate: $\frac{1}{1\times 2}+\frac{1}{2\times 3}+\frac{1}{3\times 4}+\ldots+\frac{1}{2010\times 2011}$.

2. Find the sum of the digits in the product $\underbrace{99.........99}_{\text{repeated 2011 times}} \times \underbrace{55.........55}_{\text{repeated 2011 times}}$

3. If 2 is added to the numerator of a fraction, the result is $\frac{4}{7}$; if 2 is subtracted from the denominator of the fraction, the result is $\frac{14}{25}$. Find the fraction.

4. a is a prime number and b is an odd positive number. If $a^2 + b = 2011$, what is the value of $a + b + 2$?

5. How many consecutive zeros are there at the end of 30! ?

6. The sum of the reciprocals of three prime numbers is $\frac{1155}{2006}$. Find the greatest number of the three primes.

7. A new positive integer is obtained by inserting one "0" between a 2-digit number. This new number is 9 times the original number. Find the original 2-digit number.

8. The numerator is 12 less than the denominator in a fraction. After simplification, the fraction equals $\frac{7}{13}$. Find the fraction in its original form.

9. At most how many regions can be divided by 10 lines on a plane?

10. How many squares are there in the figure?

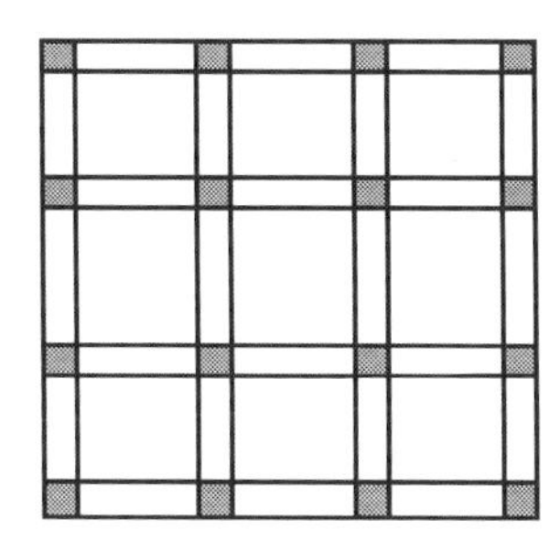

11. A 5-digit number N is written on a card. When Alex turns the card around, he sees another 5-digit number which is 78633 larger than N. Find N.

12. Coach Mark is picking up his teams of Little League Baseball. 2011 kids are lined up to be picked. His way is simple. He counts the heads starting from left to right by only saying three numbers: 1, 2, 3; 1, 2, 3;.... Then he counts the players from right to left also saying 1, 2, 3; 1, 2, 3;..... Any one who is counted "1" twice will be the team member. How many lucky kids will be picked?

13. Boxes A, B, and C contain 55 pieces of candy altogether. Box A contains three times as much candy as Box B contains. Box C contains least number of candies. Find the greatest possible number of candies in Box A.

14. There are 361 machines to be transported from River South to River North. At least how many boats are needed such that each boat transports the same number of machines and all machines are transported at the same time? It is known that no boat is big enough to hold 361 machines at once.

15. There are two water containers A and B. Container A has water weighted 0.2kg. If 1/3 of water in container B is poured into container A, the weights of water in two containers will be the same. Find the weight of water originally in container B. Express your answer as a decimal to the nearest tenth.

16. The circumference is $5\frac{5}{12}$ meters for the front tire and $6\frac{1}{3}$ meters for the back tire of a car. When the car travels n meters, the number of rotations of the front tire is 99 times more than the number of rotation of the back tire. What is n?

17. Alex, Bob, and Charles each can complete a job alone in 10 hours, 15 hours, and 20 hours, respectively. At first, Alex and Bob worked on the job for 2 hours and then Alex left. Three hours later Charles joined Bob (who did not take any breaks) and they finished the job. What is the total number of hours that the three people worked on the job?

18. Ben plans to finish reading a storybook in a certain number of days. If he reads 3 pages each day, he will leave 15 pages unfinished. If he reads 5 pages each day, he will only have 2 pages to read in the last day. How many pages are there in the book?

19. Alex and Bob are partners in a business. The sum of their money was $150,000. Last week Alex made 20% profit and Bob lost 20%. At this time, Bob's money was 1/3 of Alex's money. What was the amount of money Bob had after his loss?

20. A pharmacist wants to dilute a 20% hydrogen peroxide solution to 15%. How much 5% hydrogen peroxide solution must he add to make 450 grams of 15% solution?

21. Alex travels from Greenville to Winterville and Bob travels from Winterville to Greenville. They start at the same time and they meet at a place 700 meters from Greenville after a while. After they reach their destination, they return right way and meet the second time at a place 400 meters from Winterville. Find the distance from Greenville to Winterville.

22. It takes 2 hours for a ship to travel forth and back between Washington and New Bern. When returning to go with the current, the speed of the ship is 8 km/hour more than the speed of the ship traveling against the current. The distance traveled in the second hour is 6 km more than the distance traveled in the first hour. What is the distance from Washington to New Bern?

23. Alex is 36 years old. His son Bob is 8 years old. How many years later Bob's age will be $\frac{5}{12}$ of Alex's age?

24. A pond is fully filled with water. There is a hole at the bottom of the pond with water coming into the pond at a constant rate. The pond can be empty with 10 pumps pumping water out of it in 20 hours and 15 pumps in 10 hours. With the same rates, how many hours does it take 25 pumps to empty the pond?

25. A toy car is traveling along a 400- meter long path. The speed of the car is changing every second as follows: 2 meters per second, 3 meters per second; 2 meters per second, 3 meters per second; and so on. How many seconds does it take for the car to complete the path?

26. Alex, Bob, Catherine, David together have \$190. If \$10 is given to Alex, \$20 is taken away from Bob, Catherine's money is doubled, and half of David's money is taken away, they will have the same amount of money. What is the original amount of money for Alex?

27. During a school football game night, the principal plans to buy two drinks for every one who is attending the game night. Each bottle of drinks costs \$1.8. There are 1194 people that night. The store allows the principal to use every 6 empty bottles to exchange for a new bottle of drinks. What is the amount of money the principle saved? Express your answer in dollars to the nearest cents.

28. As shown in the figure, ABCD is a trapezoid and BD is the diagonal. The area of ΔBDC is 10 cm^2 more than the area of ΔABD. BC + AD = 15 cm. BC – AD = 5 cm. Find the area of trapezoid ABCD.

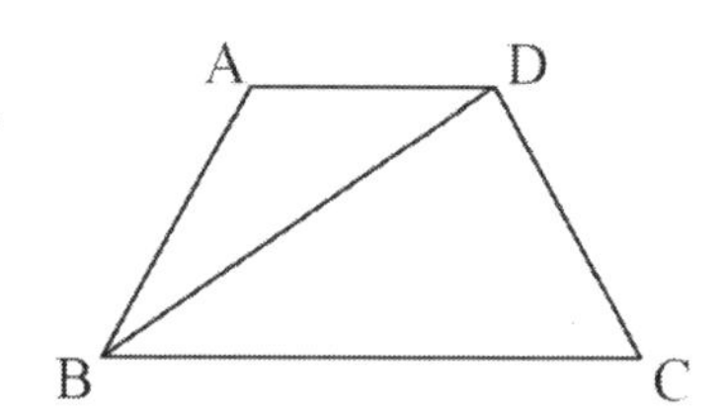

29. Shown in the figure is a portion of bee-house consisting of small hexagons. The middle small hexagon is the first layer, and the six hexagons surrounding it is the second layer. If the bee-house has a total of 6 layers and each small hexagon has a bee residing in it, how many bees are there in this bee-house?

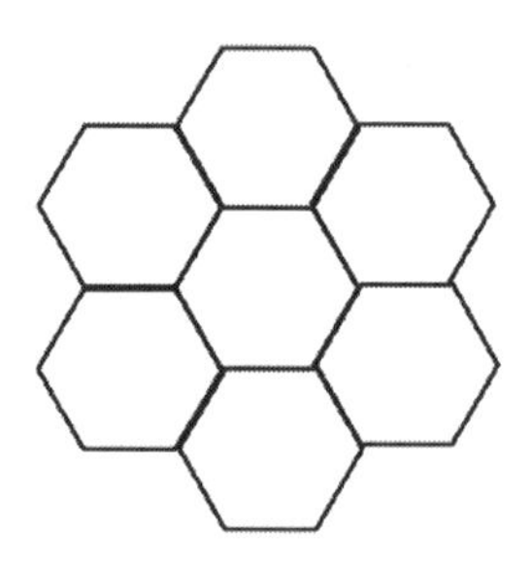

30. Using four different colors to color the five regions as shown in the figure such that neighboring regions have different colors. How many ways are there to color?

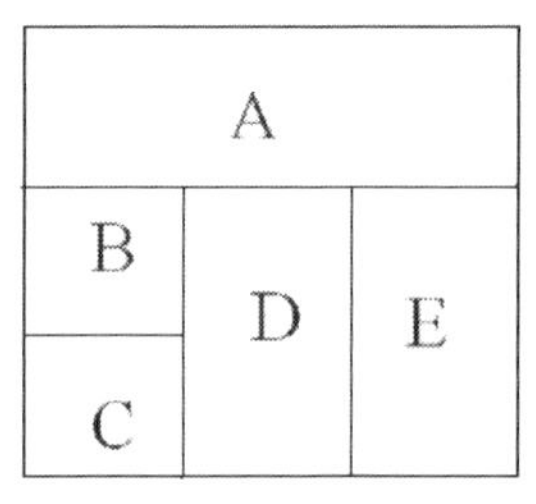

Answer keys for practice test 13

1. 2010/2011
2. 18099 (9×2011)
3. 42/77
4. 2011
5. 7
6. 59
7. 45
8. 14/26
9. 56
10. 100 ($16 + 9 + 4\times9+9+4+4\times4+4+1+1\times4+1$)
11. 10968
12. 671
13. 39
14. 19
15. 0.6
16. 3705 meters
17. 9 hours
18. 42 pages
19. $40000
20. 150 grams
21. 1700 meters
22. 15
23. 12 years
24. 5 hours
25. 160 seconds
26. $30
27. $716.4
28. 30 cm^2
29. 91
30. 96

MATHCOUNTS

■ Sprint Round Competition ■
Practice Test 14
Problems 1-30

Name

Date

DO NOT BEGIN UNTIL YOU ARE INSTRUCTED TO DO SO.

This round of the competition consists of 30 problems. You will have 40 minutes to complete the problems. You are not allowed to use calculators, books, or any other aids during this round. If you are wearing a calculator wrist watch, please give it to your proctor now. Calculations may be done on scratch paper. All answers must be complete, legible, and simplified to lowest terms. Record only final answers in the blanks in the right-hand column of the competition booklet. If you complete the problems before time is called, use the remaining time to check your answers.

Total Correct	Scorer's Initials

1. Calculate: $\left(2\frac{1}{2011}\times 3.6+3\frac{3}{5}\times 7\frac{2010}{2011}\right)\div\frac{3}{4}\div\frac{3}{5}$

2. The sum of a positive integer and its reciprocal is 20.05. Find this positive integer.

3. The sum of four positive integers is 38. Find the smallest value x and the greatest value y of the product of the four positive integers. Write your answer in the form of (x, y).

4. How many of $a-1$, a, $a+1$, and $a+2$ are composite numbers if $a=2011$?

5. p is a prime number such that both $p+2$ and $p+4$ are prime numbers. Find the value of $\frac{1}{p}+\frac{1}{p+2}+\frac{1}{p+4}$. Express your answer as a common fraction.

6. Four distinct positive even numbers a, b, c, and d satisfy that $a+b+c+d=20$. How many quadruples (a, b, c, d) are there?

7. The sum of 10 consecutive positive integers is 10055. Find the sum of the middle two positive integers.

8. The quotients are all positive integers when a number is divided by $1\frac{1}{14}$, $\frac{10}{21}$, or $\frac{20}{49}$. Find the smallest possible value for this number.

9. How many ways are there from A to B if you can only walk to left or up along the grid lines?

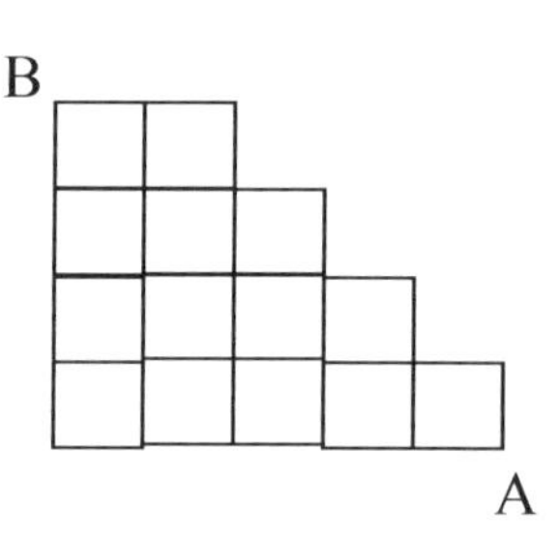

10. The numbers 1, 2, 3, 4, 5, and 6 are arranged, one per small circle, in the figure shown so that the sum, S, of the 4 numbers around each large circle is 15. What is the sum of the numbers in the two left circles?

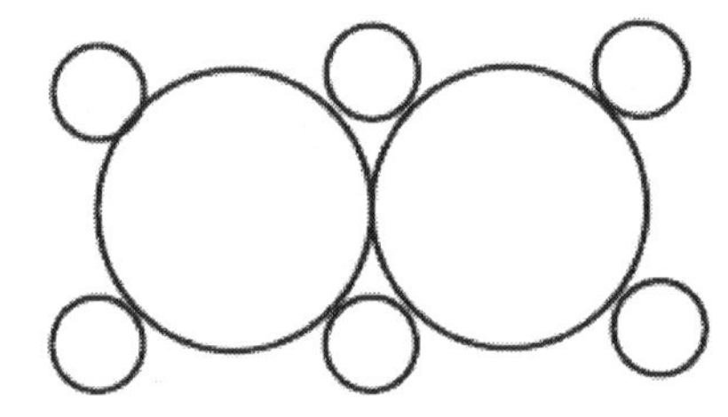

11. There are some marbles in a bag. $\frac{5}{12}$ of them are red. If 6 red marbles are added to the bag, the ratio of the number of red marbles to the total number of marbles in the bag is $\frac{1}{2}$. How many marbles are there in the bag after the addition?

12. The number of flowers that are not blooming was 55 more than 2 times of the number of flowers blooming last year in Duke's garden. There are 100 more flowers blooming this year and the number of flowers blooming is 4 times of the number of flowers not blooming. How many flowers are there in Duke's garden?

13. There are two boxes of apples weighing 110 kg together. 1/5 of apples from box A and 1/4 of apples from box B will weigh 25 kg. How many kilograms of apples are in box A originally?

14. The temperature conversion formula from Celsius (°C) to Fahrenheit (°F) is: C=(F – 32) × 5 ÷ 9. What is 32° F in Celsius °C?

15. Alex has a pile of oranges. The first time he gives away $\frac{1}{21}$ of them, the second time he gives away $\frac{1}{20}$ of the remaining, the third time he gives away $\frac{1}{19}$ of the remaining, …, and the twentieth time he gives away $\frac{1}{2}$ of the remaining. Find the ratio of the number of oranges Alex originally has to the number of oranges left.

16. Alex sold two identical shirts at $120 each. He made 20% profit on the first shirt and lost 20% on the second shirt. Find the positive difference between the money he paid and the money he made.

17. Ten kids play tennis in three courts from 1:00 pm to 5:00 pm. If each person plays the same amount of time x, what is x in hours? Express your answer in decimal form.

18. Alex has just enough money to buy only one of the following fruit: (1) 40 kg apples, (2) 60 kg oranges, or (3) 120 kg pears. In the end, he bought all three kinds of fruit with the same weight of each kind. What is the total weight of all fruit he bought?

19. One bottle contains 40% salt solution and another one contains 10%. The mixture of the two becomes 30%. If 300 grams of 20% salt solution is added, the concentration of the mixture becomes 25%. What is the weight in grams of the 40% salt solution originally?

20. Alex walks to Greenville from Washington at 9:00 a.m. at a speed of 6 km/hour. He needs to rest 15 minutes for every one-hour of walking. What is the time when he has walked a distance of 21 km?

21. The distance between two points G and Y located on Highway 94 is 4.5 km. If Alex at G and Bob at Y walk toward each other at the same time they can meet in 0.5 hours. If Alex at G walks toward Bob and Bob walks away from Alex at the same time and same direction, Alex will catch Bob in 3 hours. Find Alex's speed in km/hour. Express your answer as a decimal to the nearest hundredth.

22. Peter is walking down on an upward-moving escalator for 100 stairs to reach the bottom. Sara is walking up on the same escalator with 50 steps to reach the top. If Peter's speed is two times of Sara's speed, how many stairs are visible on the escalator when it is switched off?

23. Farmer Bob owns a piece of grassland. The grass grows at a constant rate and each cow eats at a constant rate. 17 cows can eat all of grass in 30 days. 19 cows can eat all of grass in 24 days. The height of the grass before the cows begin grazing is constant. A group of cows eat the grass for 6 days and then 4 cows are sold. The rest of cows will eat up the grass in 2 days. How many cows are there originally?

24. A nonstandard thermometer reads temperature 99 °C of a boiling water of 100 °C. It reads 4°C of a freezing water of 0 °C. It reads x °C of a classroom air of 25 °C. Find the value of x. Express your answer as a decimal to the nearest hundredth.

25. The number of 8th graders in a middle school is 36% of the total number of students. The number of 7th graders is $\frac{10}{11}$ of the number of 8th graders. The number of 6th graders is 16 less than the number of 7th graders. What is the total number of students in the school?

26. The sum of the money that Alex, Bob, and Cindy have is \$54. They want to buy a gift for their Mathcounts coach, Mr. Whitley. Each student contributes the same amount of money. Alex contributes $\frac{3}{5}$ of his money. Bob contributes $\frac{3}{4}$ of his money. Cindy contributes $\frac{2}{3}$ of her money. What is the sum of money remaining for Alex and Cindy?

27. As shown in the figure, two identical triangles have some parts overlapped. Find the shaded area.

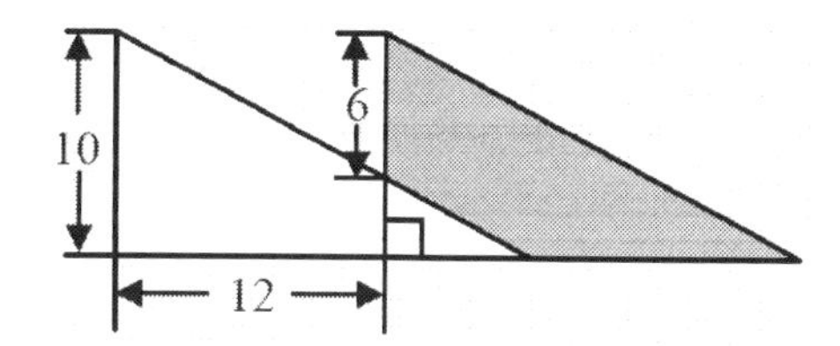

28. ABC is an isosceles triangle. D is the midpoint of the semicircle. BC is the diameter. AB=BC=10 cm. Find the shaded area in terms of π.

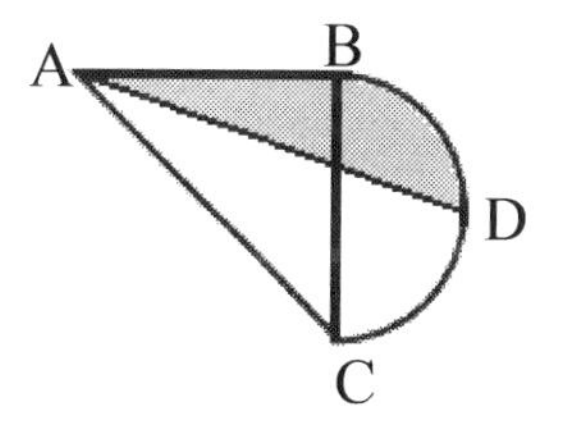

29. Three circles A. B. and C are shown in the figure. The shaded area in circle A is $\frac{1}{3}$ of the area of circle A. The shaded area in circle B is $\frac{1}{2}$ of the area of circle B. The shaded area in circle C is $\frac{1}{4}$ of the area of circle C. If the sum of the areas of circles A and B is $\frac{2}{3}$ of the area of circle C, what is the ratio of the area of circle A to the area of circle B?

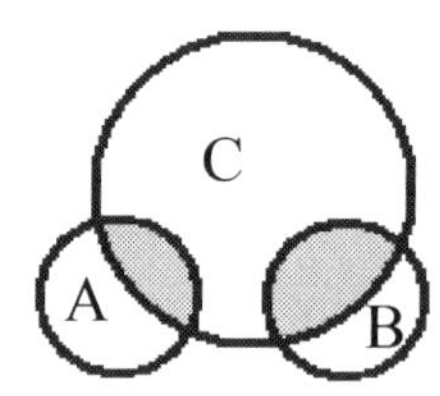

30. In a soccer tournament, the rule is as follows: the winner of a game gets 3 points, each of the two teams gets one point for a tie game, and the loser gets zero point. In the end, the champion team has the least number of wins but highest points. At least, how many points does the champion get?

Answer keys to practice test 14

1.80
2. 20
3. (35, 8100)
4. 3
5. 71/105
6. 24
7. 2011
8. 60/7
9. 90
10. 6
11. 42
12. 175
13. 50
14. 0
15. 21
16. 10
17. 2.4 hours
18. 60 kg
19. 200 grams
20. 1:15
21. 5.25
22.75
23.40
24. 27.75
25. 1100
26. 14
27. 84 cm^2.
28. $\frac{25}{2}+\frac{25}{4}\pi \text{ cm}^2$ or $\frac{1}{4}(50+25\pi)\text{ cm}^2$
29. 3:1
30.11

MATHCOUNTS

■ Sprint Round Competition ■
Practice Test 15
Problems 1-30

Name

Date

DO NOT BEGIN UNTIL YOU ARE INSTRUCTED TO DO SO.

This round of the competition consists of 30 problems. You will have 40 minutes to complete the problems. You are not allowed to use calculators, books, or any other aids during this round. If you are wearing a calculator wrist watch, please give it to your proctor now. Calculations may be done on scratch paper. All answers must be complete, legible, and simplified to lowest terms. Record only final answers in the blanks in the right-hand column of the competition booklet. If you complete the problems before time is called, use the remaining time to check your answers.

Total Correct	Scorer's Initials

1. Calculate: $(1234+2341+3412+4123)\div(1+2+3+4)$.

2. What is the greatest square number that is not exceeding 2011?

3. Using the digits 1, 2, 3, 4, 5, 6, 7, and 8 to form two four-digit numbers. What is the smallest difference of these two numbers?

4. Alex bought two horses (one black and one white) and one saddle. The white horse and the saddle cost \$800. The black horse and the saddle cost \$600. The cost for the two horses is \$1000. How much does the saddle cost?

5. A group of monkey goes out to pick up some bananas. Monkey A picks up one banana. Monkey B picks up two bananas. Monkey C picks up three bananas, and so on. When they come back home, they put all the bananas together and divide among them evenly. Each monkey gets 8 bananas. How many monkeys are there in this group?

6. Bob's class is going to a field trip to Disney World. They will rent two types of buses. Bus type A has 45 seats and the fee is \$215. Bus type B has 60 seats and the fee is \$300. If the school rents x buses of type A, there will be 15 students without seats. If the school rents $(x-1)$ buses of type B, the buses will be all full and every student gets a seat. What is the least amount of money the school needs to pay for renting the buses?

7. As shown in the figure, how many squares do not contain the letter "A"?

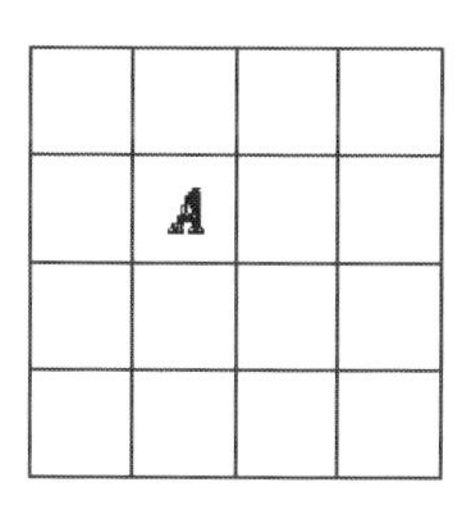

8. As shown in the figure, parallelogram ABCD is divided into a triangle ADF and a trapezoid ABCF. The difference of the areas of two shapes is 14 cm^2. AE = 7 cm. Find the length of FC.

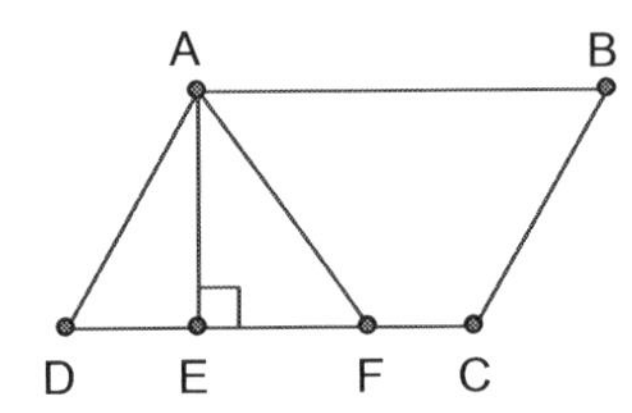

9. Catherine was reading a book. She said to Tom: "the number of pages I have read is 2.4 times the number of pages I have not read". Tom counted and said: "the number of pages you have read is 42 pages more than the number of pages you have not read". How many pages are there in the book?

10. As shown in the figure, a piece of square-shaped paper ABCD has the side length of 20 cm. Points E and F are the midpoints of AB and BC, respectively. The paper is cut along the dash lines in figure (a) and is re-arranged into figure (b). Find the shaded area.

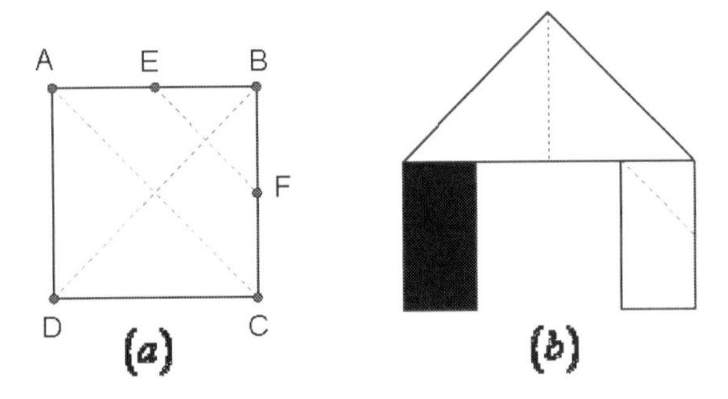

11. Alex, Bob, Charles, and Danny run for a 60-meter race. The average time for Bob, Charles, and Danny to run 60 meters is 4 minutes. The average time for Alex, Bob, Charles, and Danny to run 60 meters is 5 minutes. How long does it take Alex to run this 60-meter race?

12. A rectangular garden 345 meters long and 240 meters wide is planted with trees along the perimeter with an equal distance between any two neighboring trees. A tree is planted in each corner and at the midpoint of each side. At least how many trees are planted?

13. Alex has been throwing discus five times. If both the best score and the worst score are removed, the average score for him is 9.73 meters. If only the best score is removed, the average score for him is 9.51 meters. If only the worst score is removed, the average score for him is 9.77 meters. Find the difference between his best score and his worst score. Express your answer in the decimal form to the nearest hundredth.

14. Alex from Ayden and Bob from Beaufort are driving toward each other at the same time. They meet the first time at a point that is 32 km away from Ayden. After Alex arrives at Beaufort and Bob arrives at Ayden, they turn around right away. They meet the second time at a point that is 64 km away from Ayden. Find the distance from Ayden to Beaufort.

15. First, fold a piece of square paper along the diagonal. Second, cut a hole near each vertex of the triangle obtained. Third, unfold the paper. Which figure can you have?

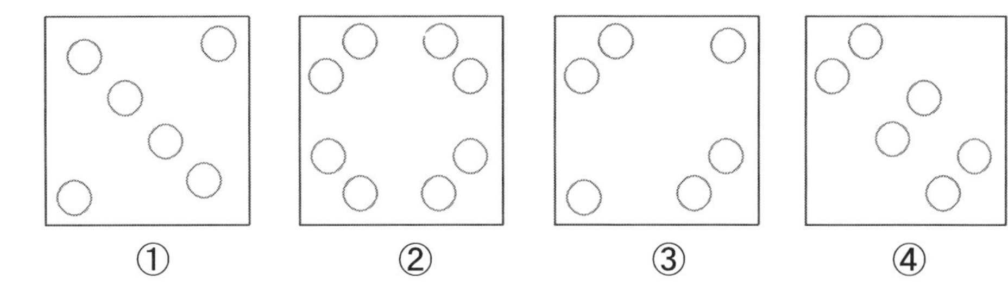

① ② ③ ④

16. As shown in the figure is a chart showing the number of cars sold in March 2010 at a car dealer store. It is predicted that the number of cars being sold in April will be 5%, 10%, and 20% increases for Toyota, Nissan and Honda, respectively. What will be the total predicted number of cars being sold in April?

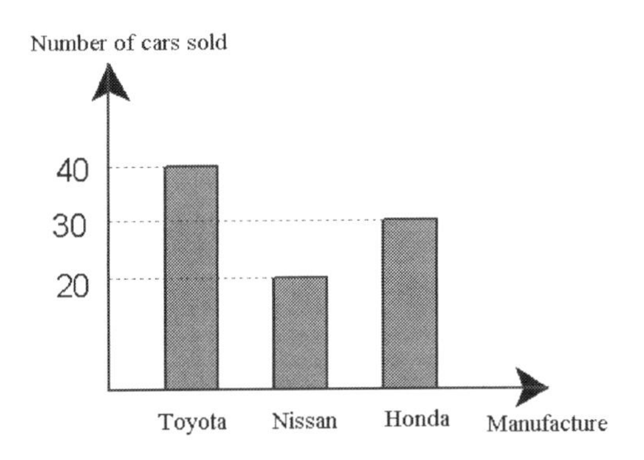

17. a and b are positive integers. Define $\otimes$ as : $a \otimes b = \dfrac{m \times a + b}{2 \times a \times b}$ (m is a fixed constant). If $1 \otimes 4 = 2 \otimes 3$, find the value of $3 \otimes 4$. Express your answer as a common fraction.

18. Find the integer part of S if $S = \dfrac{1}{\frac{1}{2005} + \frac{1}{2006} + \frac{1}{2007} + \frac{1}{2008}}$.

19. Glue 16 identical small unit cubes together to form a large rectangular prism with the volume of 16 cm^3. This large rectangular prism is then painted with red over the surfaces and then is separated back into 16 small cubes. The greatest number of small cubes with exactly three faces painted is x and the smallest number of small cubes with exactly three faces painted is y. Find the value of $x + y$.

20、The product of n counting numbers is 2010. The sum of these n counting numbers is also 2010. What is the greatest value of n?

21. As shown in the figure, there are two roads AE and CF in a piece of triangle-shaped land ABC. The intersection point is D. $DF = DC$ and $AD = 2DE$. Find the ratio of the areas of ΔACF and ΔCFB.

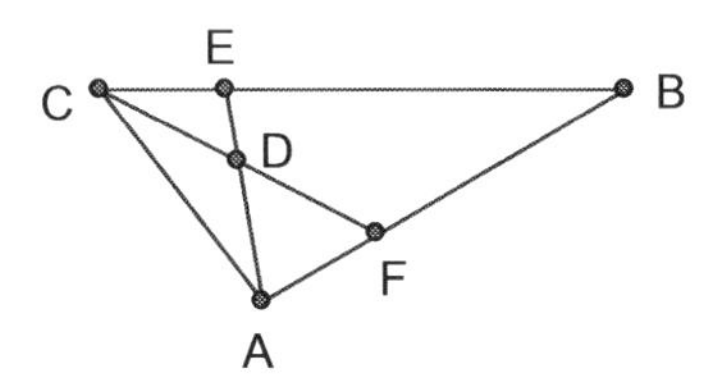

22. Two trains A and B are traveling toward each other along the parallel trails. Train A is 280 meters long and train B is 385 meters long. Alex is sitting on the train A and finds that it takes 11 seconds for train B to pass. Bob is sitting on the train B and finds that it takes x seconds for train A to pass. Find the value of x.

23. The line segments a, b, c, d, e, and f share the same vertex A. $a \perp d$, $b \perp e$, $c \perp f$. The angle formed by a and b is 30°. The angle formed by e and f is 45°. Find the angle formed by c and d.

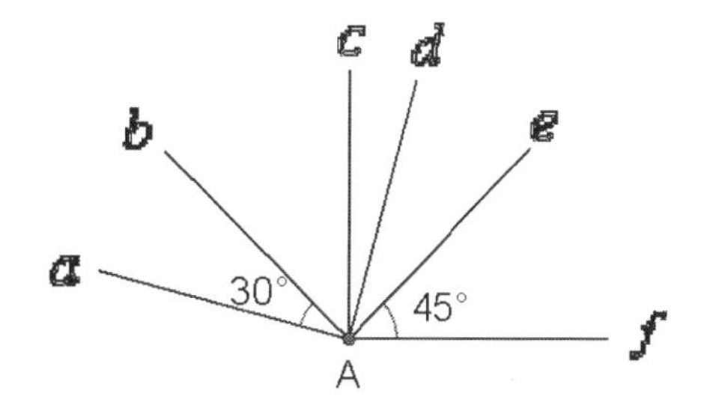

24. A special container is used to grow some microorganisms. The procedures are: first put a number of microorganisms into the container, and then seal the container. When a night is passed, the number of microorganisms is doubled. However, if the container is opened during the daytime, the number of microorganisms in the container will be reduced by 16 immediately. Mary put x microorganisms into the container in the first day. She was eager to see the test result and opened once the container during the daytime in the second day, third day, and the fourth day. On the fifth day, when she opened the container, she was surprised to find that no one microorganism was left in the container. Find the value of x.

25. Mr. Chan exercises everyday by walking three hours. First he walks on a flat road, second he walks uphill, and third he walks back along the same way. Mr. Chan's speed is 4 km per hour for the flat road, 3 km per hour for the uphill road, and 6 km per hour for the downhill road. Find the total distance he walks everyday.

26. A container is full of water. You are given three iron balls: large-sized ball L, medium-sized ball M, and small-sized ball S. First, you put ball S into the container. Second, you take the ball S out and put the ball M into the container. Third, you take the ball M out and put the ball L into the container. The volume of water that overflowed when you put ball S into the container is three times the volume of water that overflowed when you put ball M into the container. The volume of water that overflowed when you put ball L into the container is two times the volume of water that overflowed when you put ball S into the container. Find the ratio of the volumes of balls L, M, and S.

27. Ana, Bill, and Clare each have a whole number amount of money. Ana gives Bill and Clare enough, respectively, so that Bill and Clare double their amount. Then Bill gives Ana and Clare enough, respectively, so that Ana and Clare double their amount. Then Clare gives Ana and Bill enough, respectively, so that Ana and Bill double their amount. At the end, each person had the same amount of money. How much money did Ana have to begin with if she lost $100 during the exchange?

28. Fill 1, 2, 3, 4, 5, 6,7 and 8 each into a circle that is the vertex of a cube as shown in the figure such that the sum of the four numbers in each face is constant. Find the sum of three numbers that are in the circles that are connected by the line segments to the circle with a "1" in it.

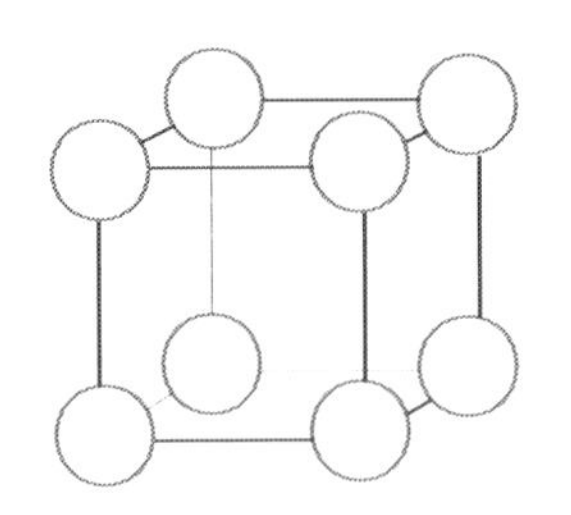

29. Water is constantly flowing into a reservoir at a rate of 40 m^3 per hour. During the first week, five water pumps were used to pump out water from the reservoir and after 2.5 hours the reservoir was empty. During the second week, eight water pumps were used to pump out water from the reservoir and after 1.5 hours the reservoir was empty. During the third week, thirteen water pumps were used to pump out water from the reservoir and after x hours the reservoir was empty. Find the value of x in hour. Express your answer in the decimal form to the nearest tenth.

30. Two roads intersect perpendicularly. Alex walks toward the North from a point that is 1200 meters South of the intersection point. Bob walks from the intersection point toward East. They start at the same time. After 10 minutes both of them are the same distance away from the intersection. After 100 minutes both of them are the same distance away from the intersection again. How far away is Alex from the intersection now?

Answer key to practice test 15

1. 1111
2. 1936
3. 247
4. 200
5. 15
6. 1160
7. 20
8. 2
9. 102
10. 100
11. 8
12. 156
13. 1.04
14. 80
15. 3
16. 100
17. 11/12
18. 501
19. 12
20. 1937
21. 1:2
22. 8
23. 15
24. 15
25. 12 kg
26. 3:4:10
27. 260
28. 21
29. 0.9 hours
30. 5400

MATHCOUNTS

■ Sprint Round Competition ■
Practice Test 16
Problems 1-30

Name

Date

DO NOT BEGIN UNTIL YOU ARE INSTRUCTED TO DO SO.

This round of the competition consists of 30 problems. You will have 40 minutes to complete the problems. You are not allowed to use calculators, books, or any other aids during this round. If you are wearing a calculator wrist watch, please give it to your proctor now. Calculations may be done on scratch paper. All answers must be complete, legible, and simplified to lowest terms. Record only final answers in the blanks in the right-hand column of the competition booklet. If you complete the problems before time is called, use the remaining time to check your answers.

Total Correct	Scorer's Initials

1. Calculate: $2^3-\left\{(-3)^4-\left[(-1)\div 2.5+2\frac{1}{4}\times(-4)\right]\div\left(24\frac{8}{15}-26\frac{8}{15}\right)\right\}$

2. Alex likes to know the time by watching the mirror on the wall. Which of the following reflection images does Alex think is closest to 8:00 A.M.?

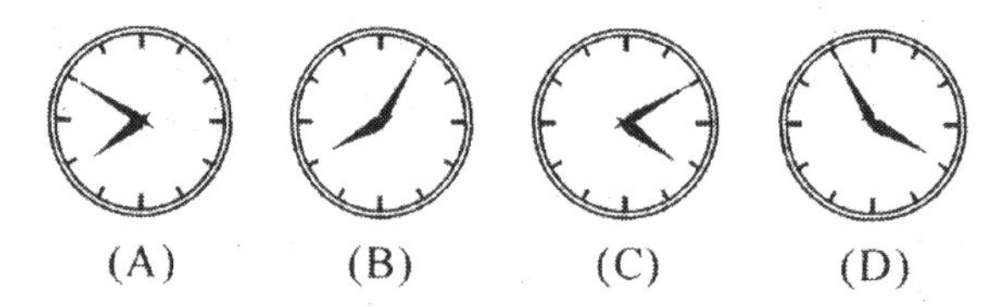

3. Sue can do $\frac{2}{5}$ of a job in $\frac{1}{5}$ hour. With this rate, how long can Sue complete the job? Express your answer as a common fraction.

4. As shown in the figure, $\angle AOB = 86°$. Ray OC bisects $\angle AOE$. Ray OD bisects $\angle BOE$. Find $\angle COD$.

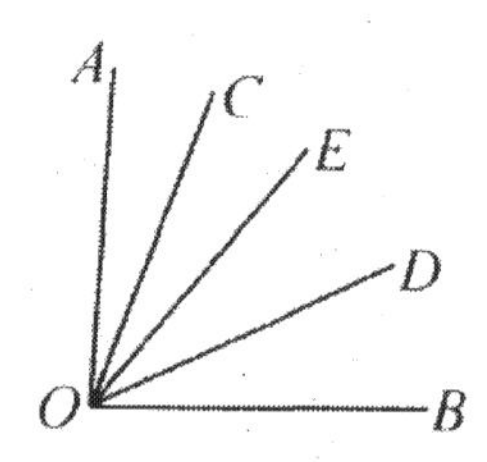

5. If $a=-1, b=0, c=1$, find the value for

$$\frac{a^{17}+b^{18}+c^{19}}{a^{20}-b^{21}+c^{22}}$$

6. In the figure, $AB \parallel CD$ and $CE \parallel FG$. If $\angle BAC = 100°$ and $\angle GFC = 110°$, find x.

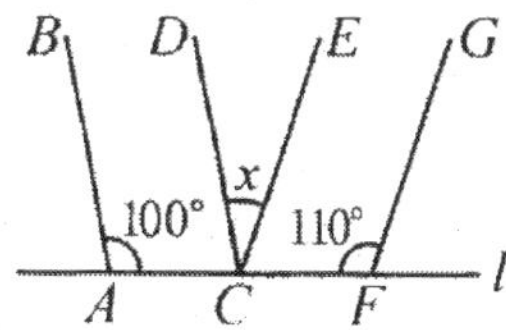

7. Simplify $p_1(x) + p_2(x) - p_3(x)$ if $p_1(x) = 2x^2 + x - 3$; $p_2(x) = x^2 - 3x + 1$; $p_3(x) = 5x^2 - 2x - 8$.

Express your answer in terms of x.

8. Find the value for y.

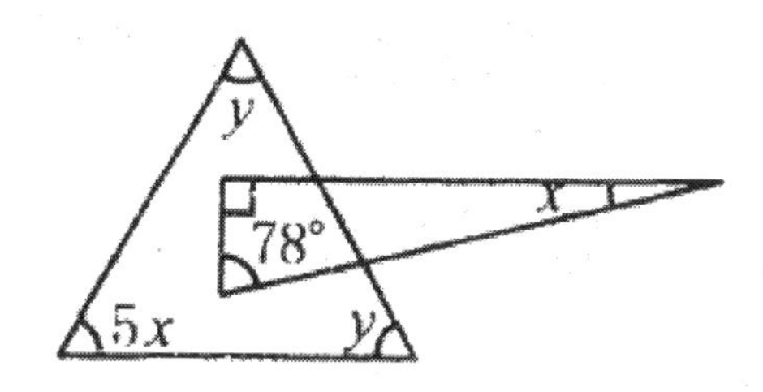

9. If the equations $2x + 7 = 3$ and $bx - 10 = -2$ have the same solution for x, what is b?

10. m is the number of triangles in figure 1 and n is the number of triangles in figure 2. Find $m + n$.

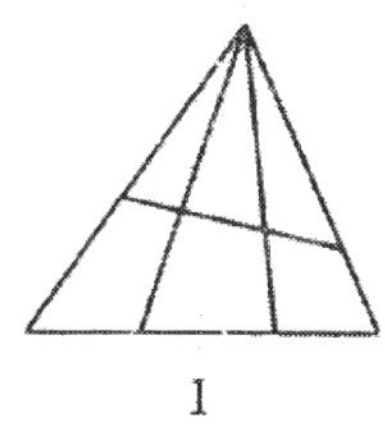
1

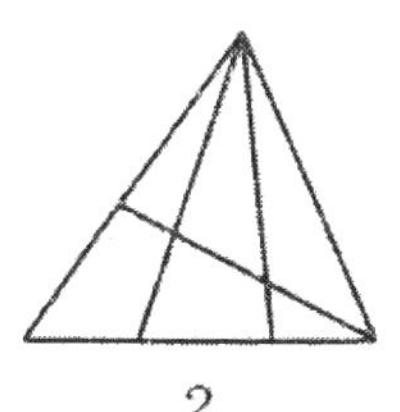
2

11. March 18, 2011 is Friday. What day was January 1st, 2008?

12. ABCD is a trapezoid with AD//BC. P is the midpoint of the diagonal BD. If the area of $\Delta APD = 15$ and the area of $\Delta BPC = 20$, find the area for ABCD.

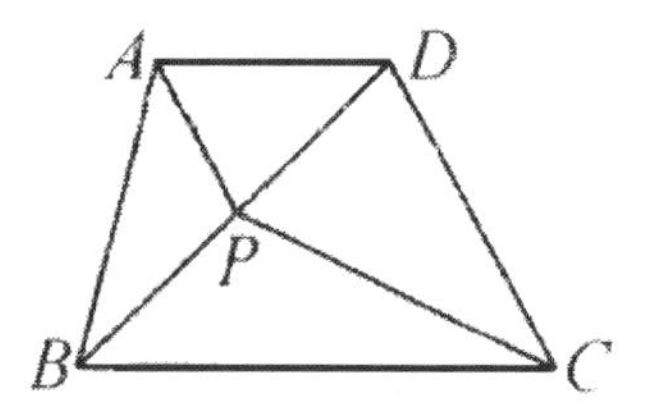

13. Find the number of solutions to the equation $2+|x-1|-|x+3|=0$.

14. Belk store's manager Alex bought a shirt at $\$a$. He first marked the sale price that was $m\%$ higher than what he paid. Later on he changed his mind and sold this shirt to his friend Bob at the price that is $n\%$ of the first marked price. What price did Bob pay? Express your answer in terms of *a, m* and *n*.

15. As shown in the figure, the smaller square has the side length of 1 and the larger square has the side length of 2. Both squares are in the same level. The smaller square now is moving through the larger square. The time used for the smaller square to go through the larger square is t_0. S represents the shaded area that is changed with time t. Which figure shows the correct relationship between S and t?

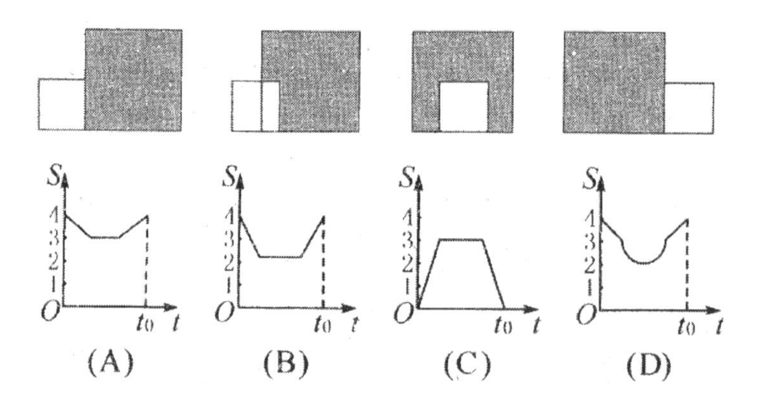

16. Square ABCD is rotated 90° along the point D in a clockwise direction. What is the sum of the coordinates of B after rotation?

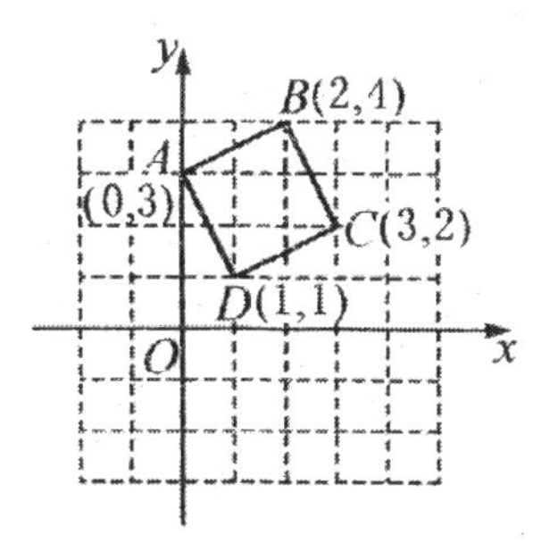

17. If the sum of reciprocals of three prime numbers is $\frac{551}{2431}$, find the sum of these three prime numbers.

18. Find the sum of the coefficients of the expansion of $(2x^2 - 7xy + 3y^2)^3$.

19. 6 rays are drawn from the point O such that OF bisects $\angle BOC$, OE bisects $\angle AOD$, $\angle AOB = 120°$, $\angle EOF = 135°$, Find $\angle COD$

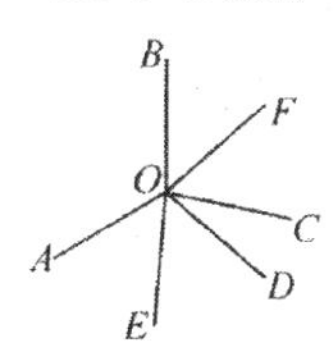

20. A solid is formed by small cubes. The different views are shown in the figure below. At least how many cubes are necessary to form such a solid?

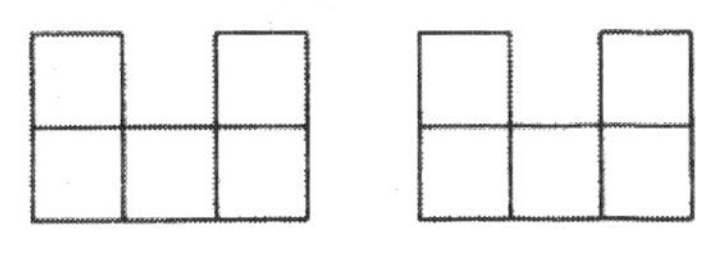

Front view Left view

21. Find the value of $x^2 - y^2 + x + y + 31$ if $x(x+1) = 3540$, and $y(y-1) = 1560$.

22. If x is a prime number, find the integer solution for the system equation

$$\begin{cases} x + y - z = 0, \\ 4x - y - z = 0 \end{cases}$$

23. 1.5 hens lay 1.5 eggs in 1.5 days. How many eggs are laid by 6 hens in 8 days?

24. Find the positive value of $a + b + c$ if

$$\begin{cases} b^2 + 2ac = 14 \\ c^2 + 2ab = 29 \\ a^2 + 2bc = 21 \end{cases}$$

25. If $\sum_{k=1}^{n} k = 1 + 2 + 3 + \cdots + n$, and $n! = 1 \times 2 \times 3 \times \cdots \times n$,

find the value of $\dfrac{2009!}{2008!} + \sum_{k=1}^{2008} k - \sum_{k=1}^{2009} k$.

26. Find the greatest value of $a^2 - ab + b^2$ if a and b are positive integers.

$2a + 3b = 15$,

27. The sum of 11 consecutive positive integers is 363. The smallest integer in the group is m and the greatest integer is n. Find $m + n$.

28. What is the smallest value of $2011(b-a)$? a and b are positive integers. $a < 2011$, $b > 2011$.

29. What is the sum of all possible values for k if $\frac{b+c}{a} = \frac{c+a}{b} = \frac{a+b}{c} = k$?

30. Find the greatest sum of A and B if $\frac{1}{A} - \frac{1}{B} = \frac{1}{988}$. Both A and B are 3-digit integers.

Answer keys to practice test 16

1. -68.3
2. D
3. ½
4. 43
5. 0
6. 30°
7. $-2x^2+6$
8. 60°
9. -4
10. 12+15=27
11. Tuesday (366 + 365 +365+77 = 4 mod 7)
12. 70. ($2\times(S_{\Delta APD} + S_{\Delta BPC})$
13. 1 (x = 0)
14. a(1+m%)n%
15. A
16. 4+0 = 4

17.41 (11, 13, 17)

18.-8

19. 30
20. 5

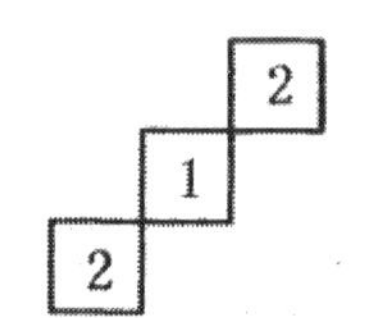

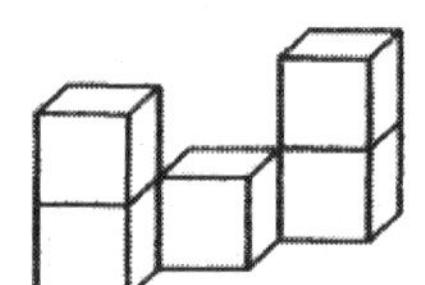

21. 2011
22. (2,3,5)
23. 32
24. 8
25. 0
26. 31
27. 28+38=66
28. 4022.
29. 1 (-1 + 2).
30. 494+988=1482

MATHCOUNTS

■ Sprint Round Competition ■
Practice Test 17
Problems 1-30

Name

Date

DO NOT BEGIN UNTIL YOU ARE INSTRUCTED TO DO SO.

This round of the competition consists of 30 problems. You will have 40 minutes to complete the problems. You are not allowed to use calculators, books, or any other aids during this round. If you are wearing a calculator wrist watch, please give it to your proctor now. Calculations may be done on scratch paper. All answers must be complete, legible, and simplified to lowest terms. Record only final answers in the blanks in the right-hand column of the competition booklet. If you complete the problems before time is called, use the remaining time to check your answers.

Total Correct	Scorer's Initials

1. Which figure, if viewed from the front, side or top, does not show any view shaped like a rectangle?

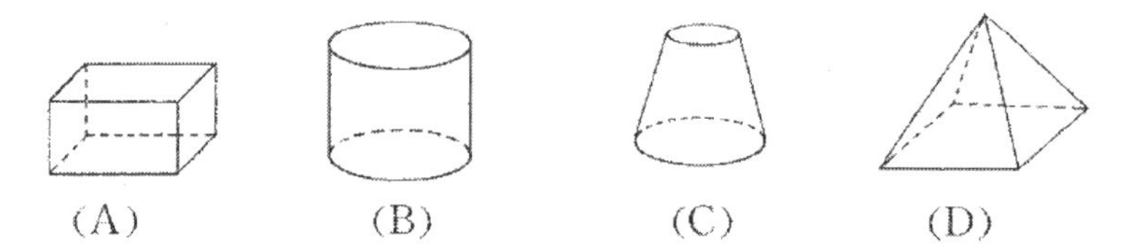

2. Find the sum of all negative factors of 105.

3. Which of the following nets cannot be folded into a cube?

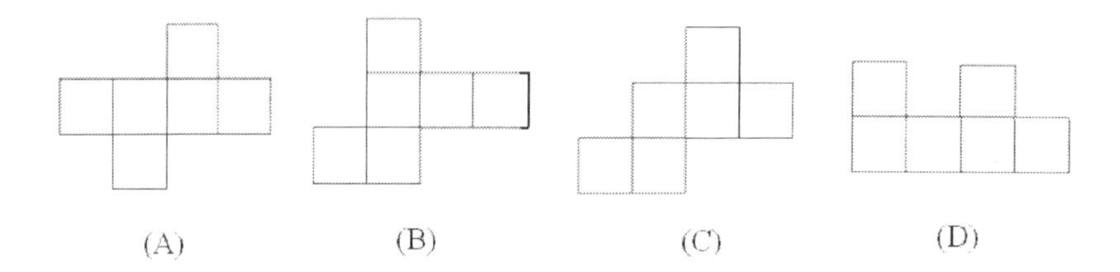

4. City workers want to pave five straight roads inside the Elm Street Park. A lamppost will be installed in any intersection of the five roads. At most how many lampposts are needed?

5. A ship moves 40 miles from A to B as shown in the figure. It then moves 40 miles from B to C. Find the distance from A to C.

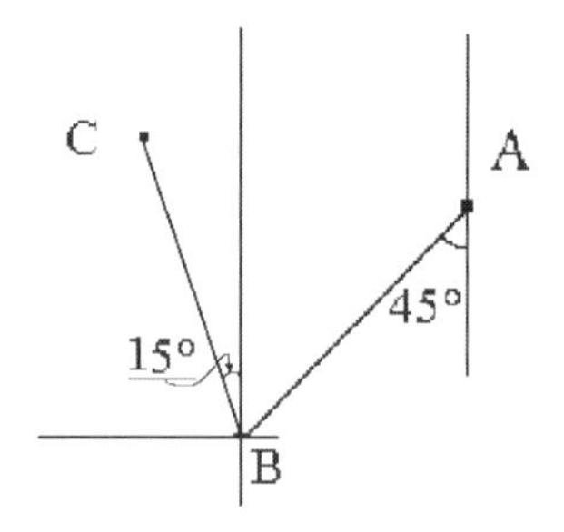

6. Calculate : $\left[-\frac{7}{5}\times\left(-2\frac{1}{2}\right)-1\right]\div 9\div\frac{1}{(-0.75)^2}-\left|2+\left(-\frac{1}{2}\right)^3\times 5^2\right|$.

7. The hard drive of a computer consists of three pie-shaped sections. The number of total spaces in each section and the rate of usage of the spaces in each section are shown in the figure. What is the total rate of usage of the hard drive?

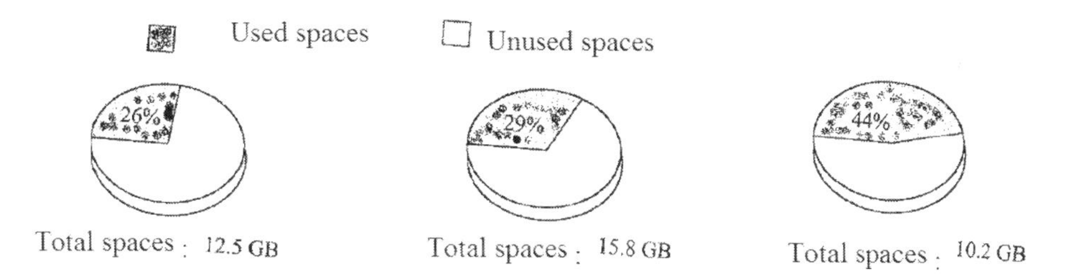

8. As shown in the figure, there are two smaller squares in a large square. The areas of two smaller squares are *m* and *n* respectively. Find *m*/*n*.

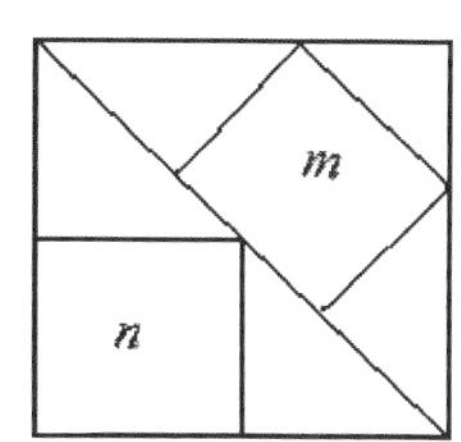

9. The large rectangle has the length of 8 and width of 6. The small rectangle has the length of 4 and width of 3 as shown in the figure. The shaded section is rotated along the dashed line to form a solid. What is the surface area of the solid?

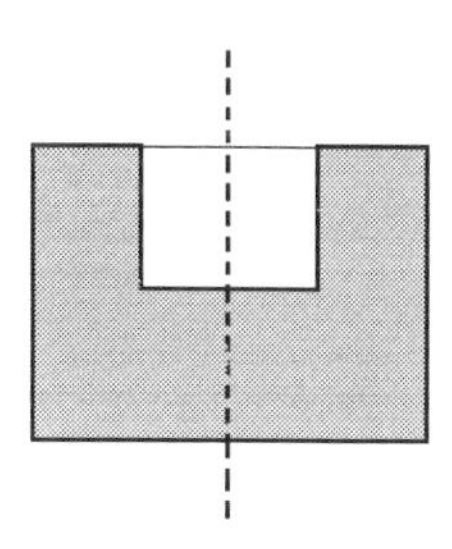

10. 120 students are assigned into 48 groups. Each group has either three girls or two boys. How many girls are there?

11. **If** $\frac{a}{b}=\frac{b}{c}=\frac{c}{a}$ **and** $abc \neq 0$, **find** $\frac{3a+2b+c}{a-2b-3c}$.

12. Alex and Bob made a trip to a food store together to buy food on two consecutive days. The food prices were different on both days. Each time Alex bought 10,000 kg food and Bob spent \$10,000. Was the total average price lower for Alex or for Bob?

13. A toy car costs \$5 and a toy truck costs \$7. Alex spent \$142 on them. What is the minimum number of toys Alex could have purchased?

14. There are m fractions between –2/3 and –1/3 with 21 as the denominator. There are n fractions between –2/3 and –1/3 with 10 as the denominator. Find the value of $m+n$.

15. Point A is –3 and point B is –1/2 on a number line. Now move the line segment AB along the number line to the right to A'B'. The midpoint of A'B' is 3. What distance did AB move?

16. As shown in the figure, each circle has the area of 30. The overlapping areas of A and B, B and C, and C and A are 6, 8, and 5, respectively. The total area covered by the three circles is 73. Find the shaded area.

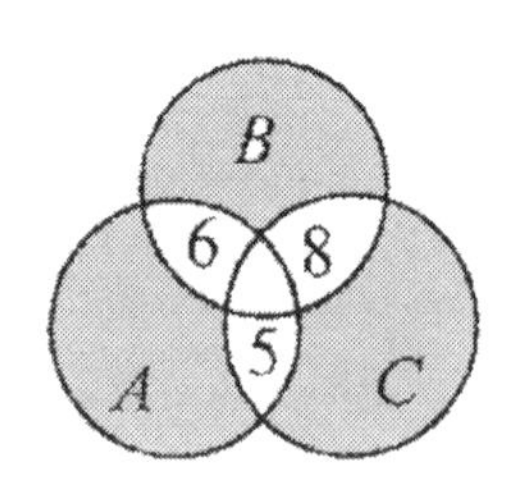

17. As shown in the figure, $\angle BOD = 45°$. Find the sum of all angles that is not greater than 90°.

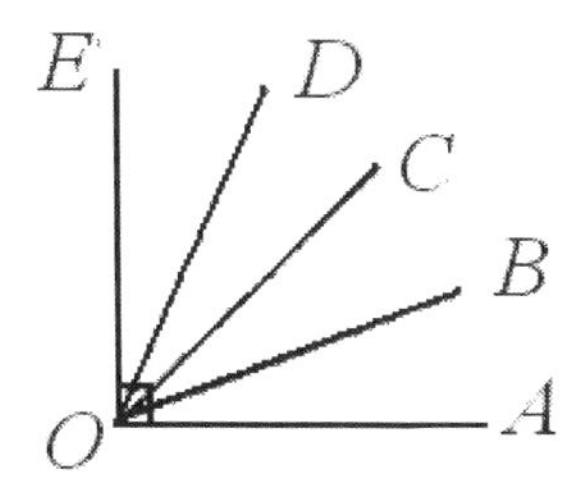

18. The fraction 6/7 is written in the decimal form. In how many digits after the decimal point is the sum of the digits exactly 2011?

19. If $(a+b)^2-(a-b)^2=4$, which of the following statements is true?

(A) a = - b. (B). a = b (C). a = 1/b. (D) a = - 1/b.

20. $x=-\frac{7}{12}$. Find the value of $(x-2)^2-2(2-2x)-(1+x)(1-x)$.

21. As shown in the figure, a solid consisting of more than 8 cubes is viewed from front and top. Which one of the following figures (A), (B), (C), and (D) cannot be obtained from left view of the solid?

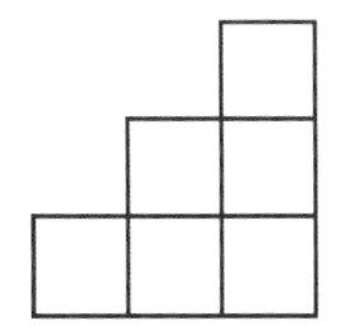
Front view

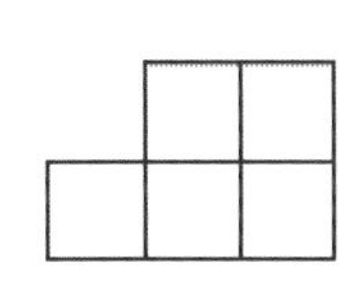
Top view

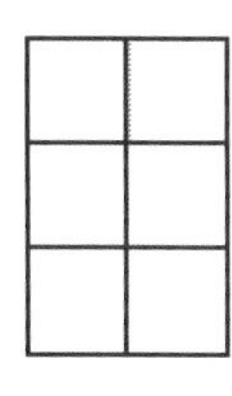
(A)

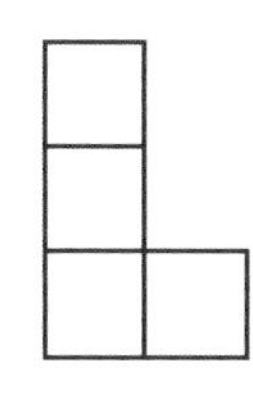
(B)

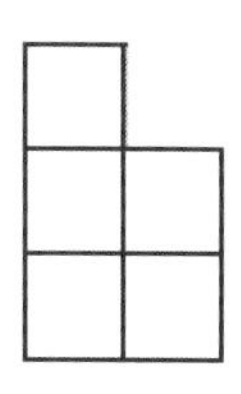
(C)

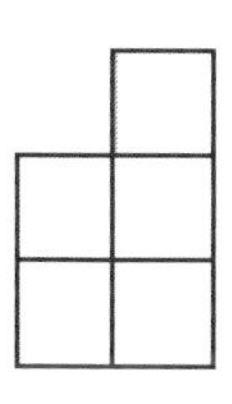
(D)

22. As shown in the figure, rectangle ABCD with $AE = BG = BF = \frac{1}{2}AD = \frac{1}{3}AB = 2$. E, H, and G are on the same line. Find the shaded area.

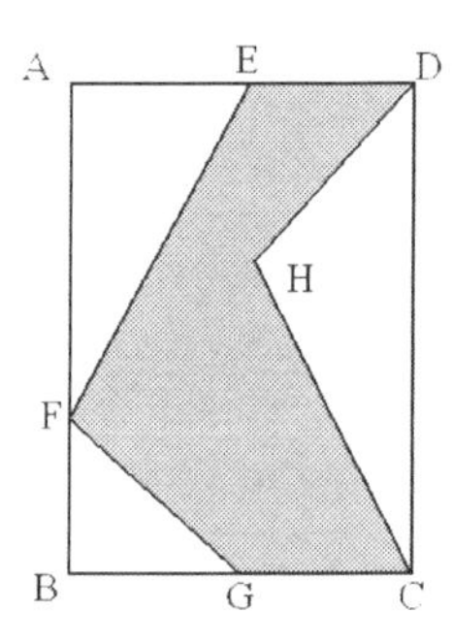

23. The measures of three interior angles in a triangle are x, $2x$, and $3x$, respectively. Find the greatest angle of the triangle.

24. Find the number of solutions to the equation $|x-2|+|x+3|=6$.

25. Which of the following inequalities is true if $|a^3-b^3|=-|a|^3+b^3$?

(A) $a>b$. (B) $a<b$. (C) $a \geqslant b$. (D) $a \leqslant b$.

26. As shwon in the figure, two lines AB and CD are parallel.

Find $\angle 1+\angle 2+\angle 3+\angle 4+\angle 5+\angle 6$

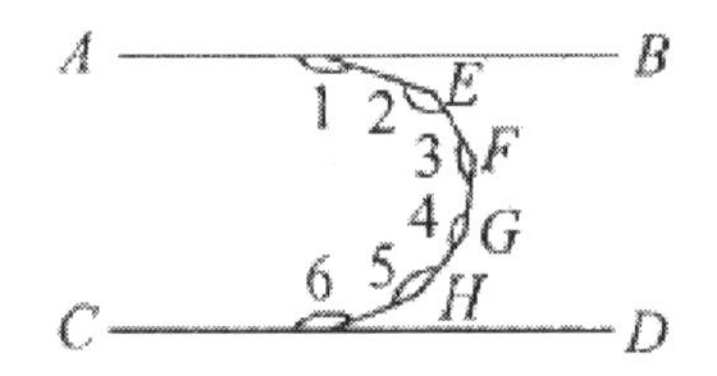

27. Find a if $x = 0.7$ is the solution to the equation: $ax + \frac{1}{2} = \frac{5}{3}$

28. Alex wanted to make n identical parts. He used machine A to make ½ of the parts and used machine B to finish the rest of the parts. He spent 4 hours to make all these parts. Machine B can make 8 more parts than machine A in every hour. Alex made 12 more parts in the second two hours than in the first two hours. Find the value of n.

29. If $x^2 + 2x = 3$, find $x^4 + 7x^3 + 8x^2 - 13x + 15$.

30. The greatest common factor of two positive integers x and y is 4. The least common multiple of x and y is 20. Find $x^2y^2 + 3xy + 1$.

Answer key to practice test 17

1. C
2. -192
3. D
4. 10
5. 40
6. –31/32
7. 32%
8. 8/9
9. 92π
10. 72
11. –3/2
12. Bob
13. 22
14. 6+3=9
15. 19/4
16. 58
17. 450
18. 446
19. C
20. –23/72
21. B
22. 12
23. 90
24. 2
25. D
26. 900
27. 5/3
28. 60
29. 18
30. 6641

MATHCOUNTS

■ Sprint Round Competition ■
Practice Test 18
Problems 1-30

Name

Date

DO NOT BEGIN UNTIL YOU ARE INSTRUCTED TO DO SO.
This round of the competition consists of 30 problems. You will have 40 minutes to complete the problems. You are not allowed to use calculators, books, or any other aids during this round. If you are wearing a calculator wrist watch, please give it to your proctor now. Calculations may be done on scratch paper. All answers must be complete, legible, and simplified to lowest terms. Record only final answers in the blanks in the right-hand column of the competition booklet. If you complete the problems before time is called, use the remaining time to check your answers.

Total Correct	Scorer's Initials

1. The students' activities at Eastern Middle School are represented by the following pie chart. Four regions correspond to the percentage of students participating in English, Drawing, Math Clubs and other activities. What percent of the students participate in the Math Club?

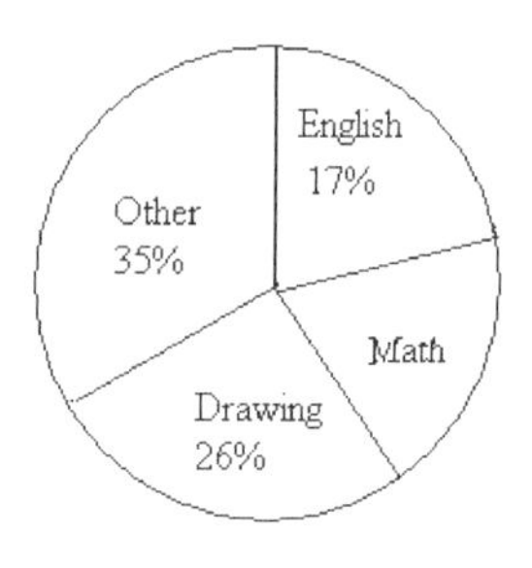

2. How many integer values of x for the equation: $\left[\frac{3x+7}{7}\right]=4$? $[x]$ represents the greatest integer not exceeding x.

3. How many positive integer solutions are there to the equation: $x+y+z=7$?

4. As shown in the figure, ABCD and BEFG are two squares. Point O is the intersection point of BF and EG. Square ABCD has the area of 9 cm^2. CG = 2 cm. Find the area of triangle DEO. Express your answer as a common fraction.

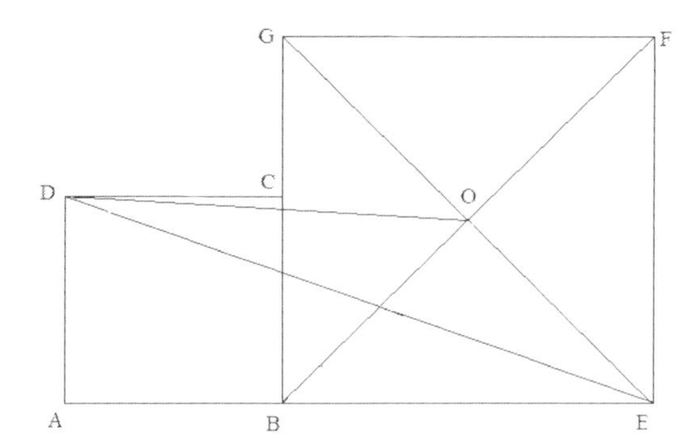

5. A rocket has the speed of 7.8 km per second. Alex, the rider, did four cartwheels in three minutes. What was the distance traveled by the rocket while Alex completed one cartwheel?

6. $a+b=3$. $a^2b+ab^2=-30$. Find the value of a^2-ab+b^2+11.

7. Ben owns some properties and the properties brought some profit for him in the last three years. Below are the profit and property value charts. Which year did Ben get the highest profit rate? Note: $profit\ \ rate=\dfrac{profit}{property\ \ value}\times 100\%$.

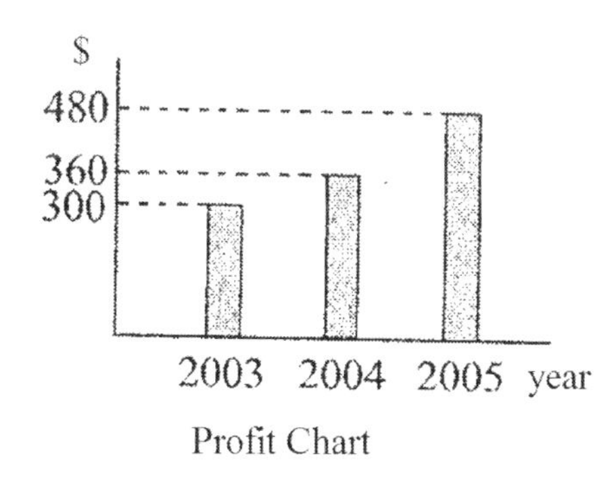

Profit Chart

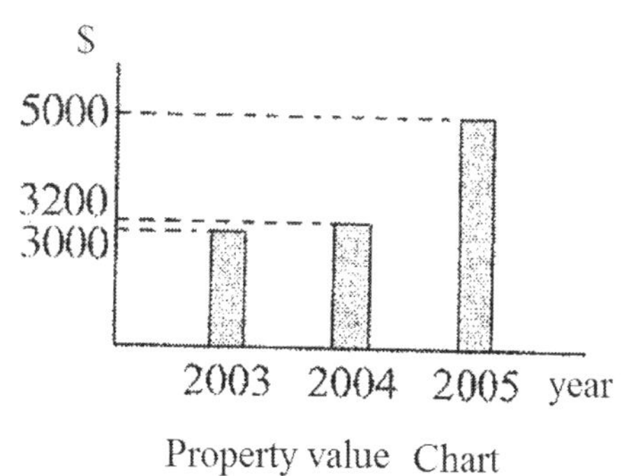

Property value Chart

8. Calculate: $$\frac{13\times 17\times\left(-\frac{2}{13}+0.125\right)\div\left(-1\frac{1}{16}\right)}{1-\frac{1}{2}-\frac{1}{8}}$$

9. The reciprocal of (m – 2) is $-\frac{1}{4}(\frac{1}{m}+2)$. Find the value of $m^2+\frac{1}{m^2}$. Express your answer as a common fraction.

10. n is a natural number. If both $n+20$ and $n-21$ are square numbers, find the value for n.

11. Find the value of a if $x = 2$ is a solution of the following equation

$$\frac{1}{9}\left\{\frac{1}{6}\left[\frac{1}{3}\left(\frac{x+a}{2}+4\right)-7\right]+10\right\}=1$$

12. The figure shows four circles each with a radius of $2\,cm$ inscribed in a square. The shaded area is $\frac{a}{7}cm^2$, find a.

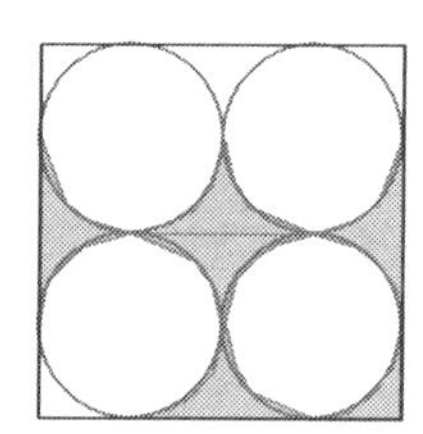

13. A tennis club sells tennis rackets with some discounts:

(a) Everyone will get a 25% discount; (b) Member of the club can get 35% more discount; (c) Senior member of the club can get 40% more discount in addition to other discounts.

If a senior member of the club bought a racket with \$585, how many dollars did he save?

14. A six-digit number $\overline{3434ab}$ is divisible by both 8 and 9. If $a+b=c$, find c.

15. A store sells some precious medal. The price is $\$0.\overline{73}$ per gram. The seller thought it was \$0.73 per gram. When he sold b kg, he found out he made a mistake and lost \$146. Find the value of b. Express your answer in the decimal form to the nearest tenth.

16. Greenville has 2476 residents. Every resident can speak English or Spanish. There are 1765 residents can speak English and 987 can speak Spanish. How many people can speak both English and Spanish?

17. If the values of x and y are decreased 25% each, what is the percent decrease of the value of xy^2? Express your answer as a common fraction.

18. As shown in the figure, DA = DB = DC. Find the value of x.

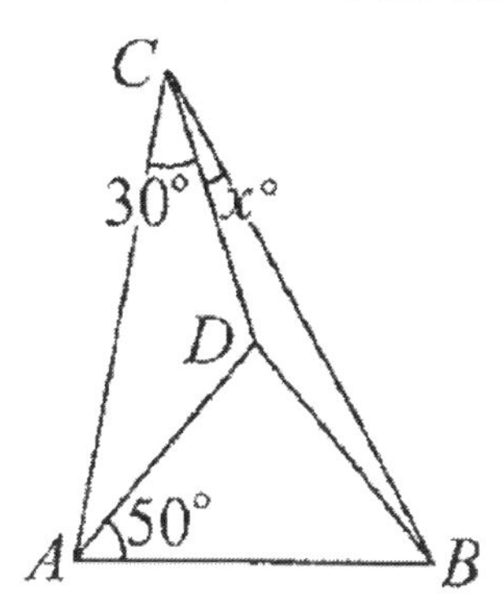

19. A 3 by 3 cube is painted red on all faces. The cube is then cut into 27 small cubes. The number of small cubes with no faces painted red is m and the number of small cubes with at least two faces painted red is n. Find $m+n$.

20. As shown in the figure, rectangle ABCD is formed by 3×4 small squares. How many rectangles are there that are not squares?

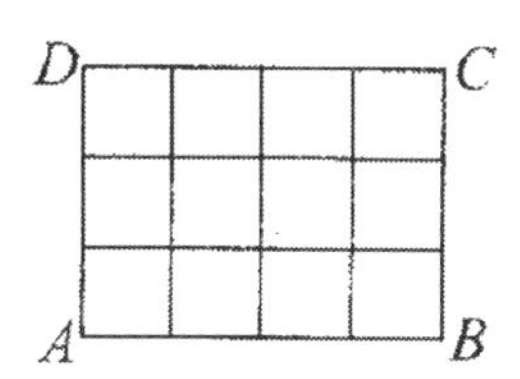

21. The circumference of the circle is 4 and is equally divided into four parts by 0, 1, 2, and 3, as shown in the figure. The number "0" on the circumference is pointing at "-1" on the number line. The circle is then rotating along the number axis without sliding. Which number on the circle will be overlapping with the number –2011 on the number line?

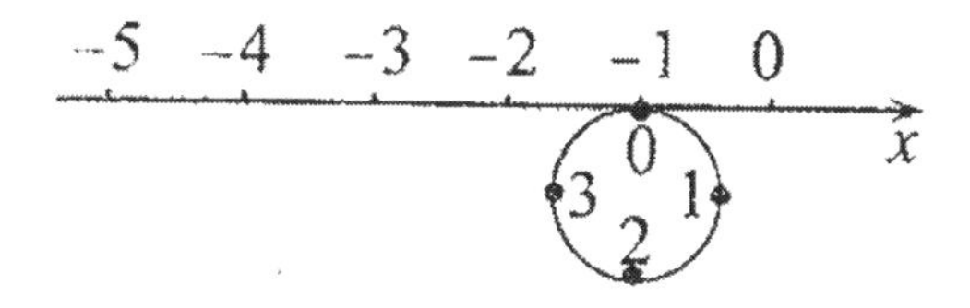

22. Triangle ABC is shown in the figure with $\angle ACB = 90°$. AC = 8 cm. BC = 6 cm. Using AC and BC as the side lengths of the two squares AEDC and BCFG, respectively, find the sum of the area of triangle BEF and the area of hexagon AEDFGB.

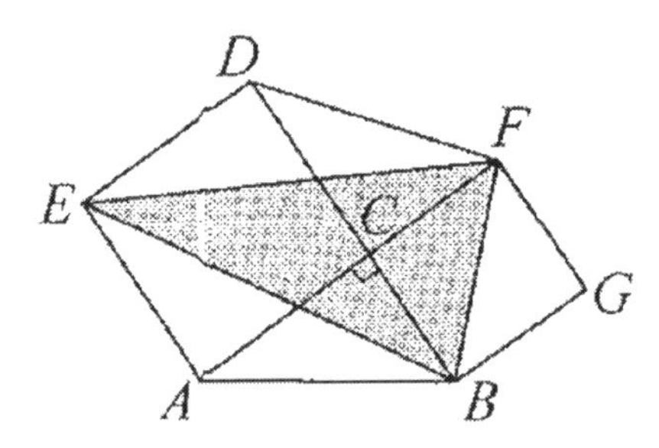

23. The three integer sides of a triangle are a, b, and c. $LCM\ (a, b, c) = 60$. $GCF\ (a, b) = 4$. $GCF\ (b, c) = 3$. Find the smallest value of $a + b + c$.

24. If m is a positive integer, what is the last digit of $2^{m+2011} - 3 \times 2^m$?

25. A street map is shown in the figure. A, B, C, …, X, Y, and Z are all intersections of the streets. A watchdog at an intersection point can watch all the streets that are connected to this point. How many watchdogs are needed in order to watch all the streets?

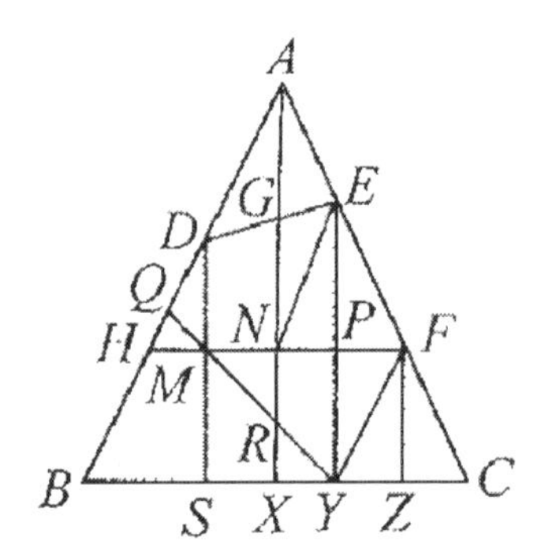

26. If $m - \dfrac{1}{m} = -3$, find the value of $m^3 - \dfrac{1}{m^3}$

27. Alex, Bob, Charlie, and Danny are friends. The average ages of any three people plus the age of the remaining person are 29, 23, 21, and 17, respectively. What is the sum of the ages of these four friends?

28. All kids in Peter's class are standing in a row for a PE class with Alex standing on the most left position and Bob standing on the most right position. Mr. Edwards starts to count the number of kids in the class. The first time he starts to count from Alex to Bob. Charles is counted "20". The second time he starts from Bob to Alex. This time Danny is counted "20". Mr. Edwards finds that there are exactly 13 kids between Charles and Danny. What is the sum of all the possible values for the number of kids in the class?

29. Alex walked from Greenville to Kinston. Emily walked from Kinston to Greenville. Alex started 5.5 minutes earlier than Emily. Emily walked 30 meters more than Alex in one minute. They met at Ayden that is between Greenville and Kinston. Alex used 4 more minutes to walk from Greenville to Ayden than the time he used to walk from Ayden to Kinston. Emily used 3 more minutes to walk from Ayden to Greenville than the time she used to walk from Kinston to Ayden. What is the distance from Greenville to Kinston?

30. Arrange 1, 2, 3, 4, 5, 6, 7, 8, and 9 in any order in a row. If any three neighboring numbers can form a three-digit number, we will get a total of seven such three-digit numbers in an arrangement. The sum of these seven three-digit numbers is N. For example, one of the arrangements is: 1, 3, 4, 2, 7, 5, 8, 9, 6. We get seven three-digit numbers: 134, 342, 427, 275, 758, 589, and 896. The sum of these numbers is $N = 3421$. Find the greatest possible value of N.

Answer keys to practice test 18

1. 22%
2. 3
3.15
4. 25/4
5. 351
6. 50
7.2004
8. 16
9. 9/4
10. 421
11.-4
12. 66
13. $1,415
14. 4
15. 19.8kg
16. 276（1765+987-2476=276）
17. 37/64
18. 10
19. 21 (20+1)
20. 40
21. 3
22. 66
23. 31
24.0
25. 4 (D, N, Y, F)
26. –36
27. 45
28. 78 (53 + 25)
29. 1440 m
30. 4648

MATHCOUNTS

■ Sprint Round Competition ■
Practice Test 19
Problems 1-30

Name

Date

DO NOT BEGIN UNTIL YOU ARE INSTRUCTED TO DO SO.

This round of the competition consists of 30 problems. You will have 40 minutes to complete the problems. You are not allowed to use calculators, books, or any other aids during this round. If you are wearing a calculator wrist watch, please give it to your proctor now. Calculations may be done on scratch paper. All answers must be complete, legible, and simplified to lowest terms. Record only final answers in the blanks in the right-hand column of the competition booklet. If you complete the problems before time is called, use the remaining time to check your answers.

Total Correct	Scorer's Initials

1. If the n-th prime number is 47, what is n?

2. Four identical dice are first put in the position (a), then are re-arranged into position (b). Find the sum of four numbers on the four bottom faces.

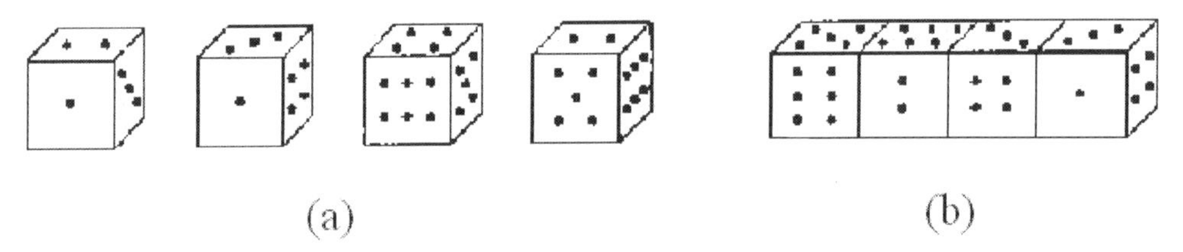

3. Define the operation of four rational numbers a, b, c, and d: $\begin{vmatrix} a & b \\ c & d \end{vmatrix} = ad - bc$. If

$\begin{vmatrix} 2x & -4 \\ x & 1 \end{vmatrix} = 18$, find the value of x.

4. Alex had 20 matches already in his tennis tournament and he won 95% of these matches. If from now on Alex always wins until the tournament is over, he will win 96% of all matches. How many matches is Alex going to play from now on?

5. As shown in the figure, D is on the leg BC of right $\triangle ABC$. BD = 2, and DC = 3. If AB = m, AD = n, find the value of $m^2 - n^2$.

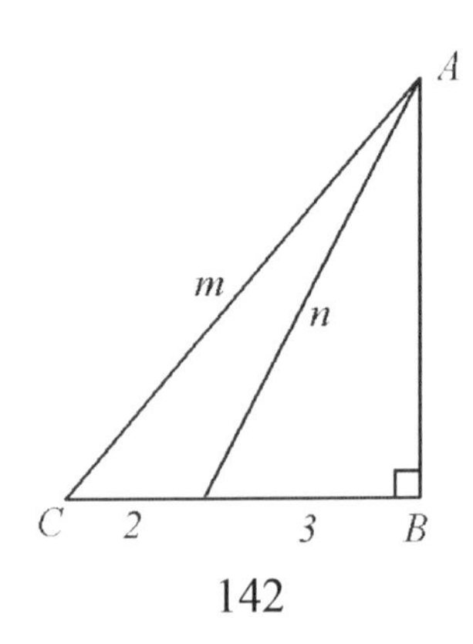

6. The average value of p, q, and r is 4, and the average value of p, q, r, and x is 5. Find x.

7. Three rational numbers a, b, and c satisfy $a : b : c = 2 : 3 : 5$. Find the value of $a+b+c$ if $a^2 + b^2 + c^2 = abc$. Express your answer as a common fraction.

8. Alex and Betsy are running along a circular path. Alex runs faster than Betsy. If they start at the same time and same location but run in opposite directions, they will meet every 25 minutes. Now they start at the same location, same time, and same direction. It takes Alex 15 minutes to catch Betsy the first time and to run a total of 16 rounds more than Betsy runs. How many rounds does Betsy run?

9. You have 100 kg of a solution that is 15% salt. You can either add x kg pure salt to bring the solution up by a 5% or add y kg pure water to reduce the solution down by a 5%. Find the value of $x+y$. Express your answer to the nearest hundredth.

10. A metal wire of 25 cm long is bent to form a triangle with three side lengths of a, b, and c with $a \le b \le c$. If a, b, and c all are prime numbers, how many such triplets are there?

11. a is a positive integer. The solution to the system equation $\begin{cases} ax+4y=8 \\ 3x+2y=6 \end{cases}$ satisfies $x>0$, $y<0$. Find the value of a.

12. The two diagonals of a quadrilateral are perpendicular and their lengths are 8 cm and 10 cm, respectively. Connect the middle points of the sides of the quadrilateral to form a new quadrilateral. Find the area of this new quadrilateral.

13. a is the length of the diagonal of a square. b and c are the lengths of two diagonals of a rhombus. If $b{:}a = a{:}c$, find the ratio of the areas of the square to the area of bthe rhombus.

14. A right triangle has one side length 11 cm. The other two sides also have integer lengths. Find the perimeter of this triangle.

15. If the lengths of three sides of a triangle can satisfy the equation $x^2 - 9x + 18 = 0$, what is the greatest possible value of the perimeter of the triangle?

16. a and b are distinct real numbers. $\frac{10a+b}{10b+a} = \frac{a+1}{b+1}$. Find the value of $a+b$.

17. Alex bought x red posters and y blue posters using \$213. Each red poster costs \$7 and each blue poster costs \$19. Find the greatest value of $x+y$.

18. Find the sum of all the integer solutions of the inequality $|x-1|+|x-2| \le 3$.

19. Point (1, 2) is on the curve determined by the function $y = \frac{a}{x}$. The function has the same y value when $x = b$ as $y = x+1$. Find the sum of all possible values of b.

20. As shown in the figure, A , B, C, and D are four small islands. Three bridges are built to connect these islands. How many different ways are there to build the bridges?

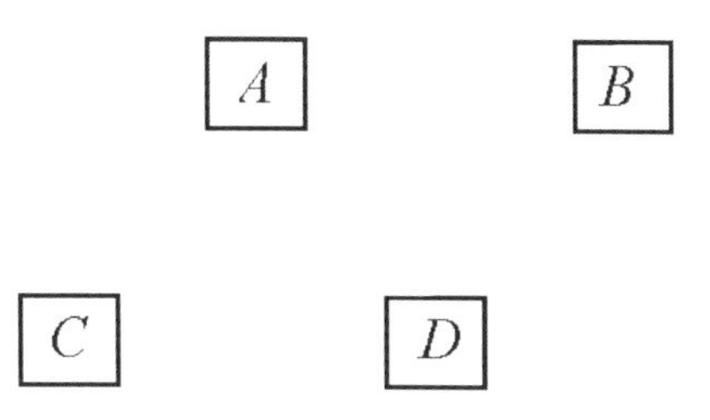

21. Three boxes A, B, and C have a total of 55 pieces of candy. The number of candy in box A is three times the number of candy in box B. Box C contains the least number of candy. At most how many pieces of candy are there in box A?

22. As shown in the figure, the number in the line segment is the time (in hours) needed to travel between the two cities the line segment connected. What is the least hours needed for Alex to travel from city A to city B?

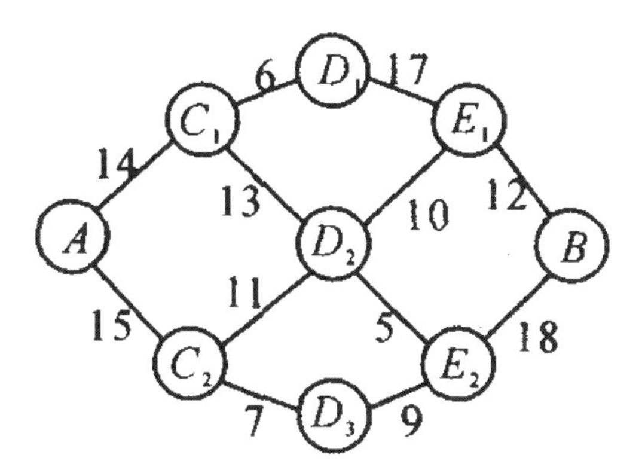

23. Five cups labeled are 1, 2, 3, 4 and 5 and five labeled lids are labeled 1, 2, 3, 4, and 5. Put five lids randomly on five cups. How many ways are there such that at least two lids match two cups with the same numbers?

24. Three boys and three girls are to be seated in a row of six chairs. Find the probability that exactly three girls are seated together. Express your answer as a common fraction.

25. The Fibonacci sequence is the sequence 1, 1, 2, 3, 5, …, where each term after the second term is the sum of the previous two terms. What is the remainder when the 2011 term in the sequence is divided by 7?

26. A time counter using tiny sands is shown in the figure. Sands are moving constantly from the upper bowl to the bottom bowl. It takes five minutes for all the sand to move from the upper bowl to the bottom bowl. H is the height of the region shown in the figure. Which figure represents the relationship between H and t correctly?

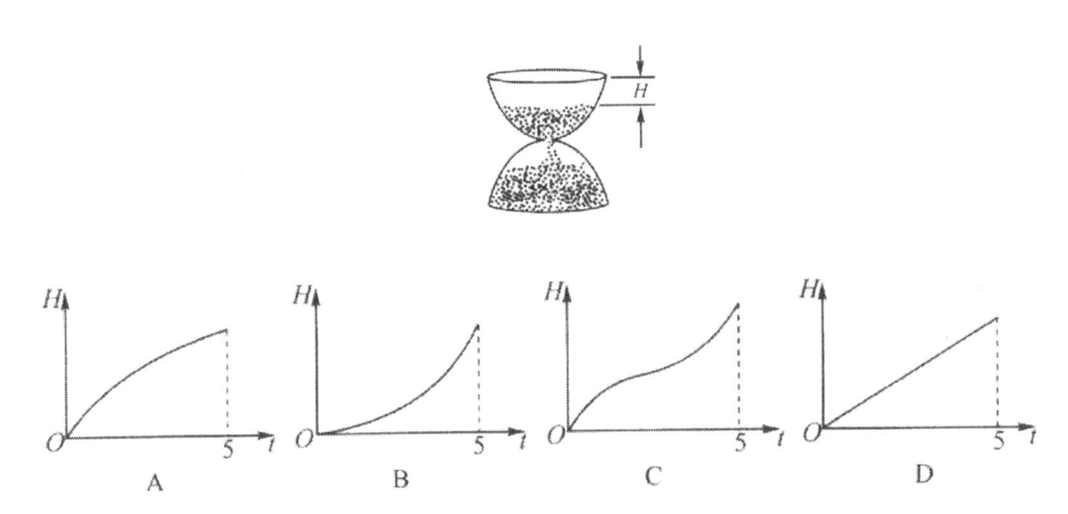

27. As shown in the figure, each switch has a probability of 1/2 to be on or off. How many ways are there for the electricity current to go from P to Q?

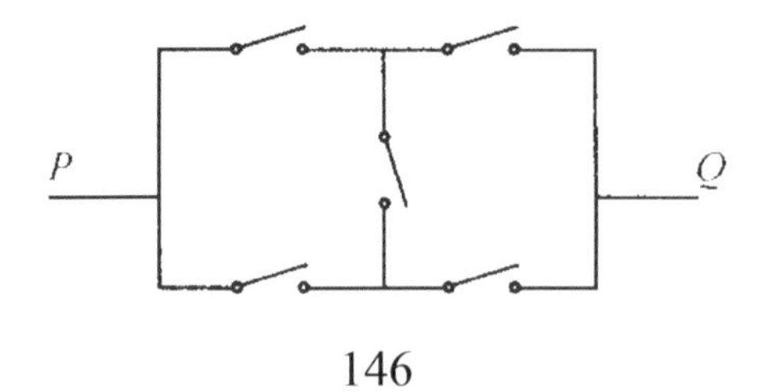

28. As shown in the figure, triangle ABC has the area of 5. AE = ED. BD = 2DC. Find the shaded area.

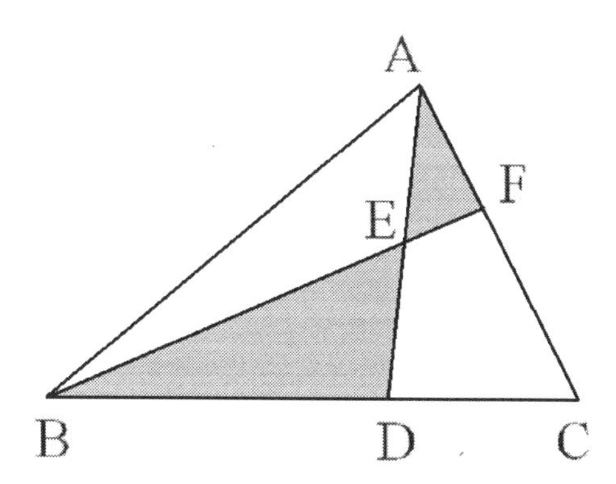

29. There is an ant located at a vertex of a regular tetrahedron. The ant walks from one vertex to another vertex along the edge with 1/3 probability of going any direction. What is the probability it comes back to the starting vertex after traveling four edges?

30. Put all natural numbers in a list from left to right as 1, 2, 3, 4, 5, ….. Remove all numbers that are multiples of 2. Remove all numbers that are multiples of 3. Keep all numbers that are multiples of 7. What is the 2011th number on the list after removals?

Answer keys to practice test 19

1. 15
2. 16
3. 3
4. 5
5. 16
6. 8
7. 38/3
8. 10
9. 56.25 (6.25 + 50)
10. 2 (11 + 11 + 13, 7 + 7 + 11)
11. 5
12. 20
13. 1
14. 132
15. 18 (3 possible values: 9, 15, and 18)
16. 9
17. 27 ($x = 2$, $y = 25$)
18. 6 (0 + 1 + 2 + 3)
19. -1 ($b = 1$ and $b = -2$)
20. 16 (12 + 4)
21. 39
22. 48 (A-C2-D2-E1-B: 15 + 11 + 10 +12)
23. 31
24. 1/5
25. 5. (The pattern repeats every 16 terms when mod 7: 1, 1, 2, 3, 5, 1, 6, 0, 6, 6, 5, 4, 2, 6, 1, 0). $2011 \equiv 16\times265+11$).
26. B
27. 16
28. 2
29. 7/27
30. 4691

MATHCOUNTS

■ Sprint Round Competition ■
Practice Test 20
Problems 1-30

Name

____

Date

____

DO NOT BEGIN UNTIL YOU ARE INSTRUCTED TO DO SO.

This round of the competition consists of 30 problems. You will have 40 minutes to complete the problems. You are not allowed to use calculators, books, or any other aids during this round. If you are wearing a calculator wrist watch, please give it to your proctor now. Calculations may be done on scratch paper. All answers must be complete, legible, and simplified to lowest terms. Record only final answers in the blanks in the right-hand column of the competition booklet. If you complete the problems before time is called, use the remaining time to check your answers.

Total Correct	Scorer's Initials

1. Calculate: $2008 \times 2006 + 2007 \times 2005 - 2007 \times 2006 - 2008 \times 2005$.

2. When a positive integer m is divided by another positive integer n, the quotient is 20 and the remainder is 8. What is the smallest possible value for m?

3. How many squares in the figure do not contain the letter A?

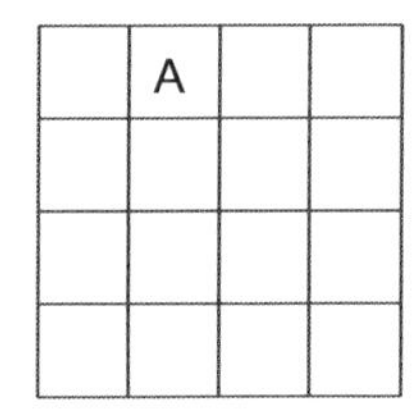

4. Alex has a book of x pages. If he reads y pages each day, he can finish the book in 20 days. If he reads y+3 pages each day, he can finish the book in 16 days. Find the value of x.

5. Alex's age is $\frac{1}{4}$ of Bob's age. After 15 years, Alex's age will be $\frac{5}{11}$ of Bob's age. How old is Alex now?

6. In a certain time interval, the sum of the numbers of complete rotations made by hour, minute, and second hands is exactly 1466. How many seconds are there in this interval?

7. Alex and his brother Bob walk to school from their home along a straight road. They start at the same time. Alex's speed is 60 meters per minute and Bob's speed is 90 meters per minute. Bob turns around right after he arrives at the school and meets Alex on the way. After they meet, Alex spends three more minutes to arrive at the school. Find the distance from Alex's home to his school.

8. As shown in the figure, ABCD is a rectangle. When a square ADFE is cut off, the perimeter of the rectangle EFCB is 100 cm. Find the length of AB.

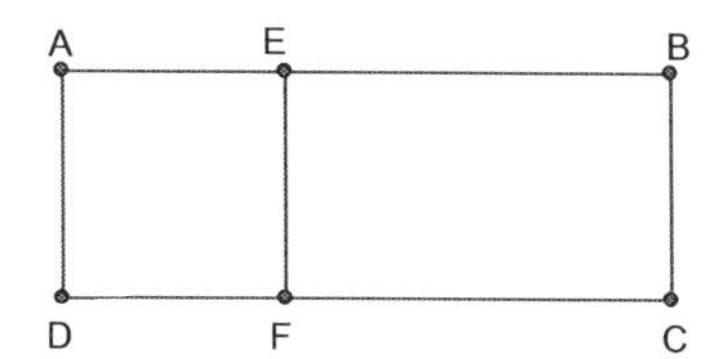

9. As shown in the figure, there are three squares and one circle. The area of the biggest square is 60 cm^2. What is the area of the smallest square?

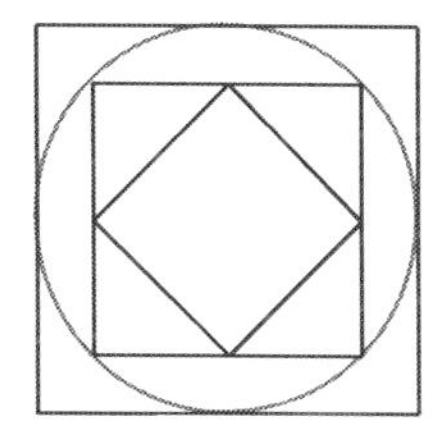

10. Three identical cubes with each face using one of the labels A, B, C, D, E, and F are shown below. Which letter is opposite to the letter E?

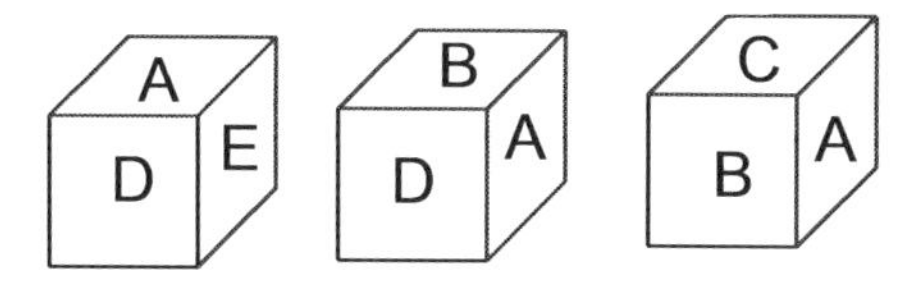

11. Sam's class made some toy pigs. Each student made a paper-pig first. Then every two students working together made a clay-pig. After that every three students working together made a cotton-pig. At last, every four students working together made a metal-pig. When they finished, Mrs. Whaley counted a total of 100 toy pigs made by the class. How many students were there in Sam's class?

12. a, b, and c are all natural numbers. Define the operation: "♣" : $♣(a, b, c) = \frac{a-b\div c}{a+b\times c}$. Find ♣(1, 2, 3). Express your answer as a common fraction.

13. When counting number n is divided by 2, the remainder is 1. When n is divided by 3, the remainder is 2. When n is divided by 4, the remainder is 1. When n is divided by 5, the remainder is 1. Find the smallest value for n.

14. What is the value of p^5+5 if both p and p^3+5 are prime numbers?

15. The following four figures are composed of simpler figures A, B, C, or D with each simpler figure a line segment or a square . The sign of the composition is written as "*".

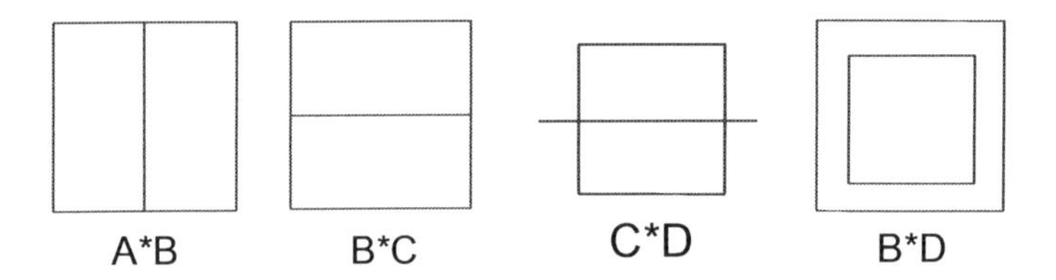

Use the letters and the composition sign to represent the following figure.

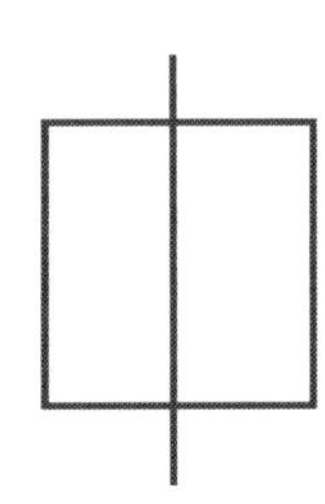

16. Three identical red marbles and four identical yellow marbles are arranged in a row such that no red marbles are next to each other. How many ways are there to do so?

17. x, y, and z are distinct natural numbers. If $\frac{1}{x}+\frac{1}{y}+\frac{1}{z}=1$, find the value x+y+z.

18. If $a:b=\frac{3}{2}:1.2$, and $b:c=0.75:\frac{1}{2}$, find the value of $c:a$. Express your answer in simplest ratio form.

19. Calculate: $\dfrac{\left(1-\frac{1}{2}\right)\left(1-\frac{1}{3}\right)\left(1-\frac{1}{4}\right)\left(1-\frac{1}{5}\right)\left(1-\frac{1}{6}\right)\left(1-\frac{1}{7}\right)\left(1-\frac{1}{8}\right)\left(1-\frac{1}{9}\right)}{0.1+0.2+0.3+0.4+0.5+0.6+0.7+0.8+0.9}$. Express your answer as a common fraction.

20. Insert the signs $+$, $-$, $\times$, or $\div$ into the small square 1□2□3□4□5 to form an expression. Each square is inserted with one sign and no sign is allowed to use twice. What is the greatest possible result?

21. As shown in the figure is a 3×3 magic square. Each square is filled with one number such that the sum of three numbers in each row, column and each diagonal is the same. Which number should go in the square marked "*" ?

3		
*		4
7		

22 Alex can finish a job in 10 days. Bob can finish the job in 15 days. Charlie can finish the job in 20 days. They worked together for three days then Alex left. Bob and Charlie continued to work together x days to finish the job. What is x?

23. A solar system contains two planets and a sun. Planet A completes an orbit around the sun in $1\frac{4}{5}$ months. Planet B completes an orbit around the sun in x months. Every 144 months, Planet A circulates 35 more rounds than the Planet B does. Find x and express it as a common fraction.

24. Three numbers p, $p+1$, and $p+3$ are all prime numbers. What is the reciprocal of the sum of the reciprocals of three numbers? Express your answer as a common fraction.

25. The number "0" is inserted into a two-digit positive integer to form a new three-digit number. The three-digit number is one less than 8 times the original two-digit number. Find the original two-digit number.

26. How many different ways are there from A to B if one can only move to the right or down?

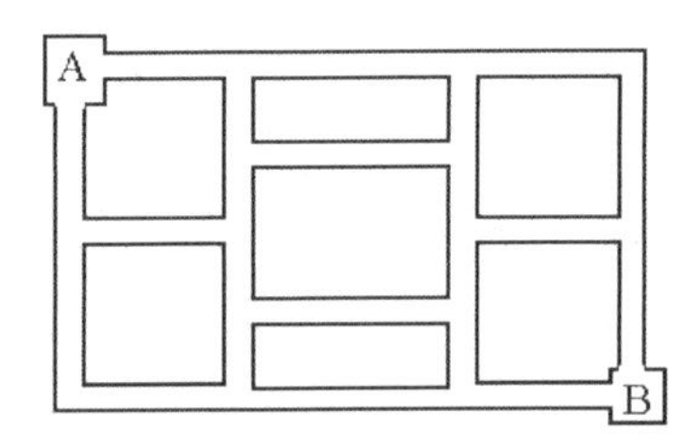

27. An electronic watch shows the time at 10 o'clock, 28 minutes and six seconds below. If we write down the time, we get a six-digit code: 102806. How many such six-digit codes can be obtained from 10 am to 10:30 am if all digits are different?

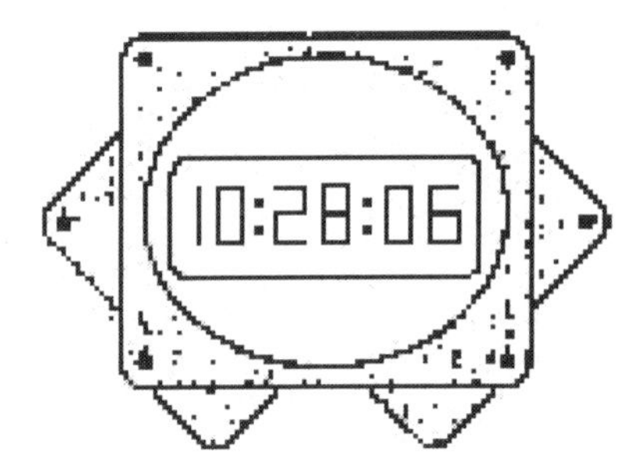

28. As shown in the figure, *ABCD* is a square with the side length 10 cm. AB is also the diameter of the half circle. Find the shaded area. Express your answer in terms of π.

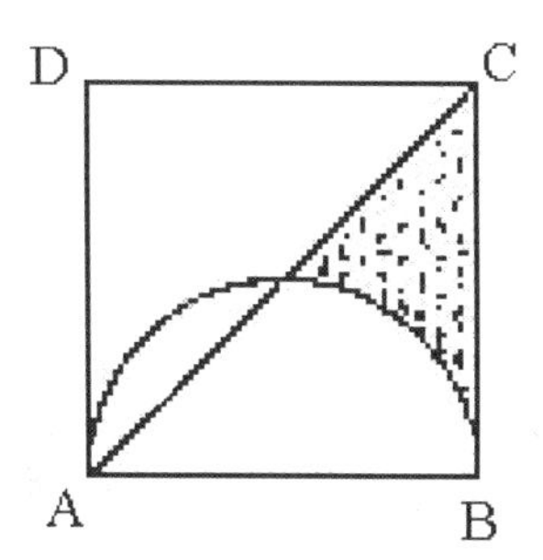

29. Alex hates banks, and he puts all his money into a box under his bed. There is some money in his box now. His salary is a fixed amount each month. He puts the money after spending in his moneybox. If he spends $1,000 each month, one and half years later his moneybox will have $8,000. If he spends $800 each month, two years later his moneybox will have $12,800. Find the amount of money now in his moneybox.

30. The salt in a cup of water becomes 15 % after a certain amount of water is added. The salt becomes 12 % after the same amount of water is added the second time. The salt becomes x % after the same amount of water is added the third time. Find the value for x.

Answer keys to practice test 20

1.	1.
2.	188
3	24
4	240
5	10
6	86400
7	900
8	50
9	15
10	B
11	48
12	1/21
13	41
14	37
15	A*D
16	10
17	11
18	8:15
19	2/81
20	61/3
21	8
22	3
23	16/3
24	30/31
25	13
26	10
27	90
28	17.875
29	8000
30	10

Made in the USA
Lexington, KY
14 November 2012